—∞∞∞—

THE LONG CYCLES PLANETS RESOLVED

planets

* OSCLILLATION OF THE EARTH
* THE TRUE SECULAR CYCLES
  AND SECONDARY CYCLES

PAUL HENRI BRESTEAU

ISBN-13: 9781511832342
ISBN-10: 1511832347
Library of Congress Control Number: 2015915783
CreateSpace Independent Publishing Platform
North Charleston, South Carolina

TABLE OF CONTENTS

PREFACE

The purpose of studies is to reconstitute the true
Trajectories Planetes Solar systeme to wit

Mercury, Venus, Mars Jupiter,

Saturne;Uranus;Neptune;pluton.

Indeed the trajectories of the planets compose themselves of cycles long told
secular and short. The difficulty cycle is to make difference among these cycles
to have a clear approach to the computations to do in the purpose to get accu-
rary average and long term.

To obtain clear view cycles the files of the observatories and the catalogs of
measuring are used by the method of the reconstituted node

OBSERVATORIES

The observatories function as références in
Positions of planets, astéroids, stars, galaxy
and multidisciplinary searching
Some of these observatories specialize themselves
In one of these categories and edit
Yearly, pluri- yearly or decennial catalogs.
Others edits series of some years
Concerning a specific element of the sky
The catalogs give RA α (right
ascension) and β (Declination) relatively
to an equinox of reference J1950 or J2000.
Services are working to record discovery
and give a name to the objects celestials.
Today, there exists more 24 telescopes
With a diameter equivalent or superior
at 4 meters.

These are classical mirrors big
diameter or well several mirrors big
Sizes of which the light

Is returned towards a central collector.
These mirrors are offsetted streaming in temperature by
Active sensors.
Electronics of management of the set
Mecanism oversees and carries out the compensations
necessery at the clarity image.
Special programms of computation analyse
are recording these images or films.
Films are also used and not only image
to have dynamics details.
The world locate of the observatories is develop
to have an atmosphere the most transparent.
The choices of the places are selected according to several criteres:
The high mountains
The deserts
The islands
These places owe farthest as possible of
Human activities
And particularly of the towns (luminous pollution)
And factory pollutant (chemical clouds)

THE CATALOGS

In U.S.A.exist a military organism of state:
USNO. USNO means United State Naval
Observatory. The american department
Collect planetary positions from 1850.
He has an important park observatories
On the ground americain who supply it
Measures planets. The astrometric department
edit several catalogs
Pluri- yearly name: series W.
3 catalogs used as references function for
Computations theoriques:
*W50
author :j. a. Hugues and d. K. Scott
Years: 1963-1971
W1 J00
Author: Rafferty And Holdenied
Years :1974-1981
W2 J00
Author :Rafferty and Holdenied
Years :1984-1996

*** exploitation Catalog
The paragraphs which interesting us
Entitle :
Six inch transit circle
Observations of the sun, moon, and
Planets and stars.
The planets do not incorporate pluto
and stop at neptune.
In page 5 Some of the catalog is shown
Like example:
Column 1 :Greewich dates (dates year+month+day)
colonne2 :Julian Ephemeris Dates
The date like Julian day
colonne3: Observatory
The name of the observatory in short
Column 4 : geographic locate of the observatory
W or E, West or east
Column 5 :Right ascension
Longitude in hours, minutes, seconds
Column 7 :O-C
Comparison between the measure of the observatory
And the americans computations in second
Column 8: Declination
Latitude In degre, minutes arc, secondes arc
Column 9 :O-C
Comparison between the measure and
american Computations in arc second.
Column 10: Distance

OBSERVATOIRIES EUROPE CATALOG

Some France it is the observatory of BORDEAU
Who carries out measures of astrométries
With CCD for numerous years.
Of the service astrometry planéts, i got
A long pluri- yearly series of 2001 at
2004 of the planet pluton. Same argument for
The planets uranus and neptune and Saturn.
Address:
Observatory of BORDEAU
2 street The observatory
bp 89
33270 Floirac
France
Site web of presentation also

There exists also the catalogs CARLSBERG,
Since the years 1980 several catalogs
Were published. Here incorporate measures
Stars, planets and astéroids. Theses measures
Are carried out by the observatory of La Palma

Situated on a island in the south Spain and
By others in Europe according to the numbers.
These catalogs are available for consultation in
Bookcases of observatory and in particular
In Paris:
Paris 's Bibliotheque
77 ave Denfert Rochereau
75014 Paris
France
To Paris there is the Bdl= the office of the longitudes
Who has a site « astrometrie database planet »
This site incorporates the resultats of measures
Several country 's observatory

CALENDAR

Scaliger start the timetable in -4712
Period told romaine. Since this date on the year -45
Numerous modifications were brought about
At level structure her months, days and month
Intercalary (mensis intercalaris). The purpose of these timetables is to
Correspond March 21 with equinoxe of springtime
Date reference for the agriculture. For this he must
Make similar year tropic (solar) 365.242219 days
And the legal year with a number in days whole.
With this end in mind that one has create the additional days
(Intercalary) who give the years bisextiles
In the year -46 Jules Cesar named Pontifex Maximus
Re- formed the calendar giving it a modern structure
Here start in the year -45. Unfortunately
The priests responsibles with the application deceived themselves
With intercalary all one day the 3 years rather than
4 ans. The emperor Auguste realize the mistake
And corrected 36 years after the using. The correcting
Made, the Julian timetable started fevrier year 4.
After several centuries of use, the Pope Gregoire XIII

Wanting to correct time different between March 21
And equinoxe springtime re- formed the Julian timetable
In calendar gregorien For that he suppressed 10 days.
One passed of October 4, 1582 at October 15, 1582.
This re- shape broadcast itself fast in Europe
And more slowly moreover the (1582 at 1927)
*Années bisextiles
Are declared bisextiles years
Divisible by 4 (bisextile)
Except the seculars which are divisible by 100 (not bisextile)
Seculars divible by 400 (bisextile)
Seculars divisible by 4000 (bisextile)
Formulae of Scaliger giving the Julian day starting with the 2 timetables
JD =fix (365.25* (an+4716)) +fix (30.6001* (mo+1)) +jo1+b+jo-1524.5

ΔT

The time whom regule our sociéties has first of all
Been based on the spin of the earth (time Solar.)
Ensues the universal time UT, civil time, was
Create and has for origin greenwich meridian
(Timezone =0.)
TU= true time + equation of the time
     + longitude (West) + N hours
TU= true time +équation of the time
     − longitude (east)− N hours
The universal time is second reference :
s=1/86400. Since 1967 second is no longer
define by the spin of the earth but by
Clocks atomic at cesium stable
In the time. The atomic second is
universal base of coordinated time UTC.
There is several UT 's variants
UT0;UT1;UT2;UTR1
UT0= he is measured of multifold manner by report
Diurnal motion stars
Radiosources extragalactic

Observations multifold of the Moon
Artificial satellites
UT1 =UT0 corrected of the motions of the poles
Which induce variations of several meters
UT1 has one uncertainty +- 3 ms by day
UT1R is a slicked down value of UT1
Short term variations are leaked.
UT2=UT1 +corrections of the irregulars
most important spin of the earth.
**UTC
Stable Scale but included time
Enter the atomic international time TAI
And the time earthly UT.
UTC=TAI- TU < 0.9 s
Thus 0.9s is maximum difference
Authorized between these 2 time, one stable and
The other variable
To respect the diference one has to correct
Earthly time cause by slowdown
Earth: 0.002 by century + attraction moon-sun.
+tide +storm
The correctnesses of the unaccounted for seconds (second leap)
Are carried out June 30 or 31 december

***the time in the ephemerides

The astronomers to compute the positions
Planetes and stars used TE

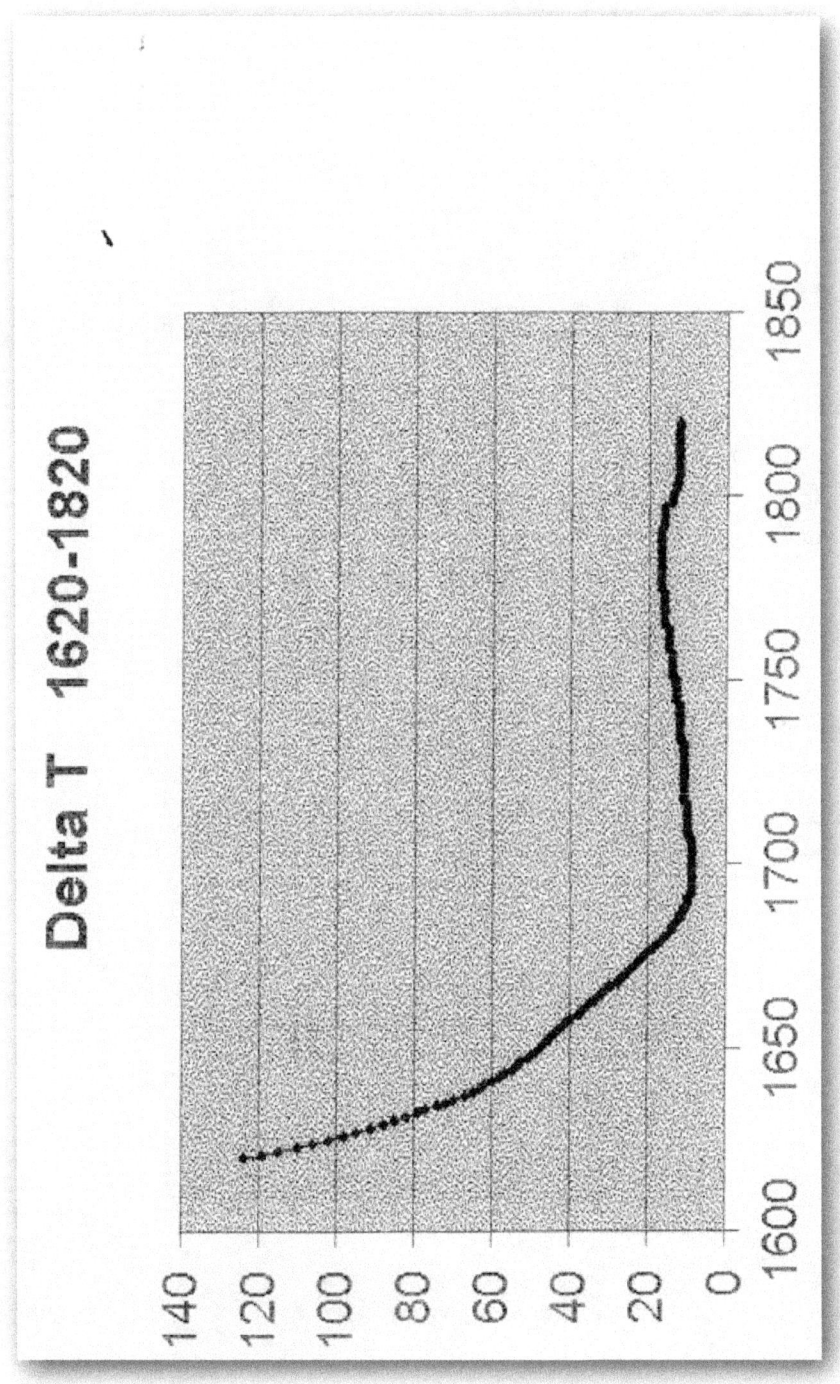

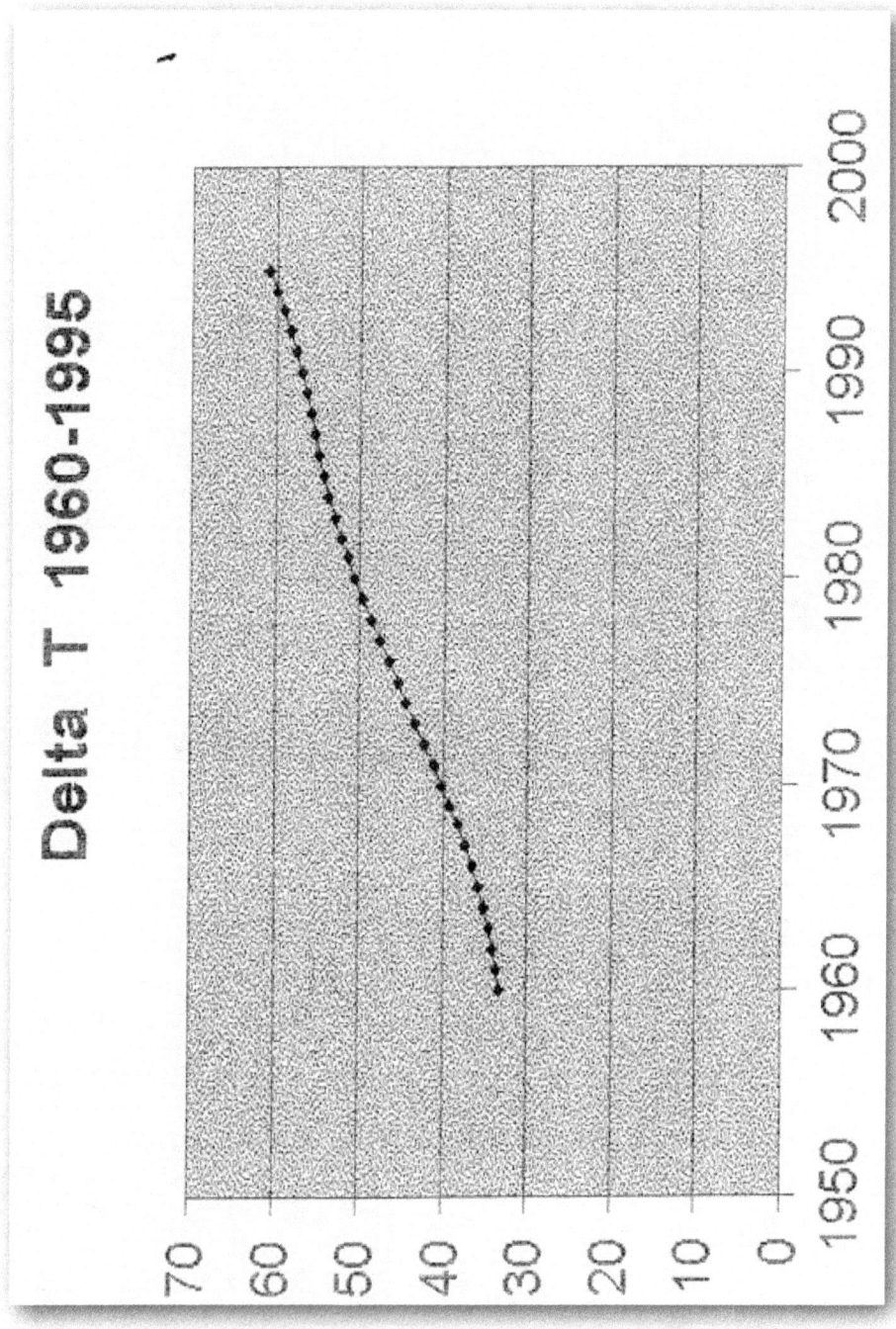

Delta T 1960-1995

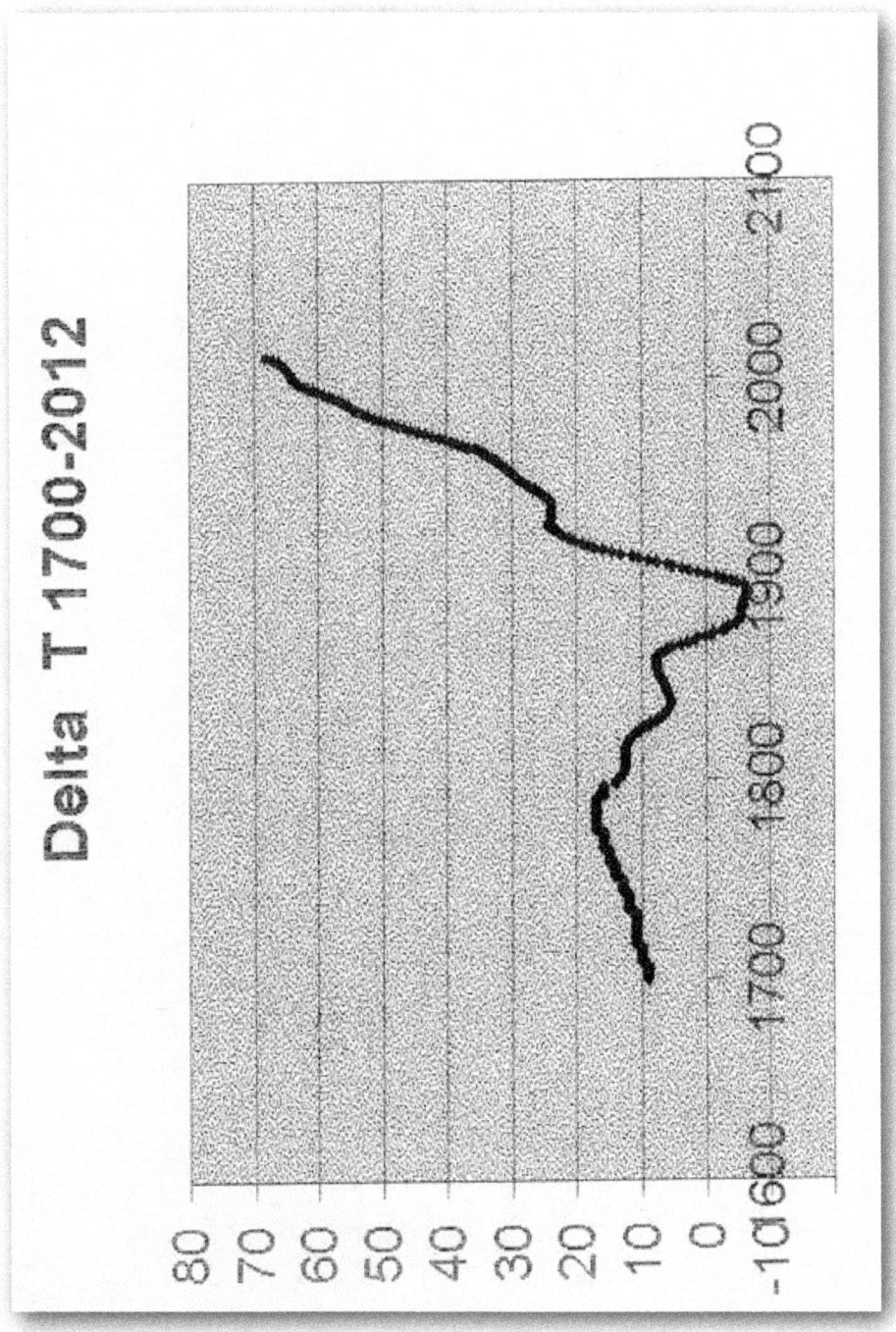

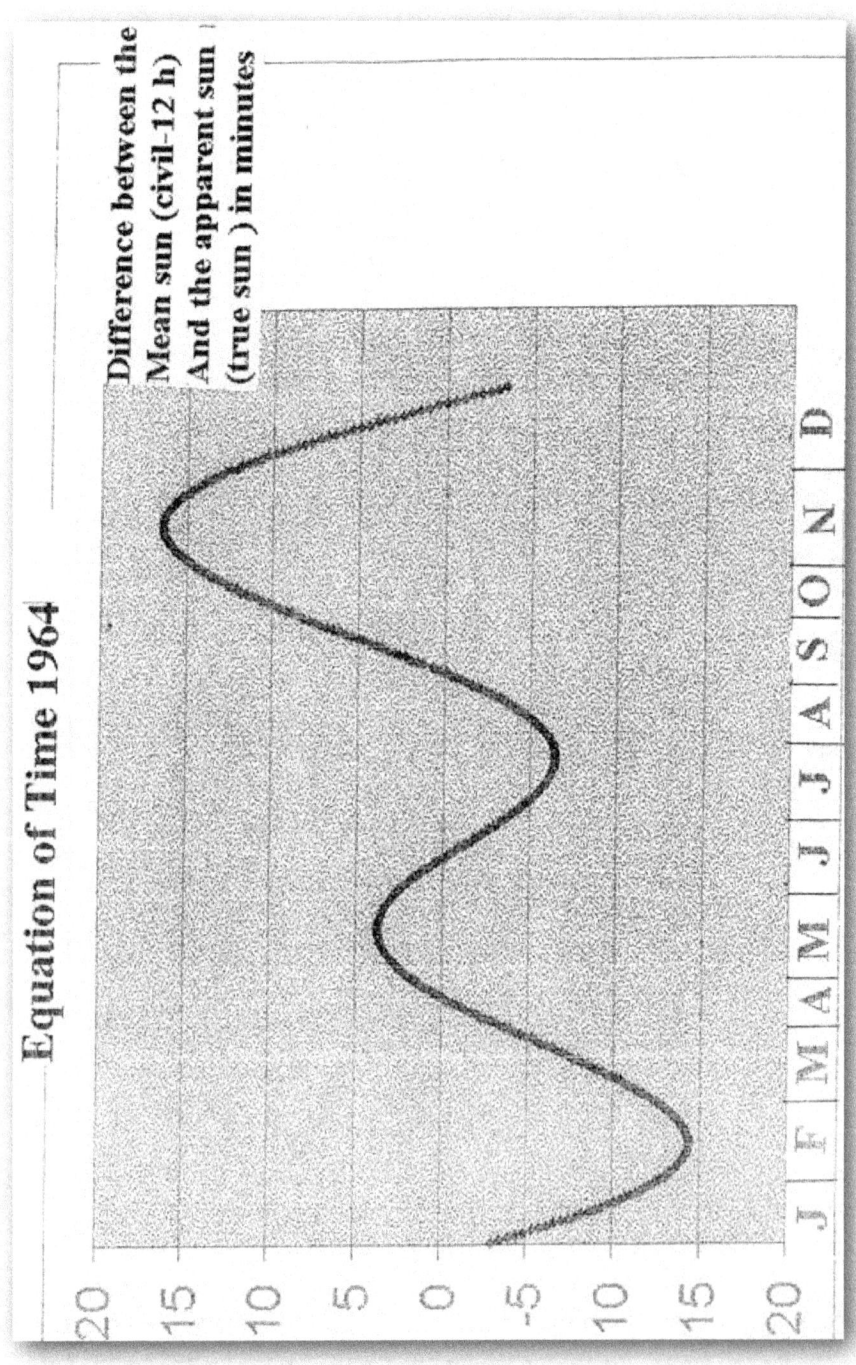

Equation of Time 1964

Difference between the
Mean sun (civil-12 h)
And the apparent sun
(true sun) in minutes

THE ORBITAL ELEMENTS

The elements orbitals of the planets present themselves
Under the polynomial shape: a+b*T+c*T*T+d*T*T*T
where T is the time in relation to a date reference.
Example:
January 1900 0.5 JJ=2415020
January 1950 0.923 JJ=2433282.423
1.5 January 2000 JJ=2451545
exemple :
epoch 2000 : formulae
T=(JDE-2451545)/36525
These elements
Are reported to average equinox of reference.
These elements are :
L=longitude average of planet
A=semimajor axis of the orbit
E=excentricity of the orbit
I=inclination on the plane ecliptic
ω =argument of perihelion
Ω =longitude of the ascending node
pi=longitude of the perihelion

(pi is a variable and don't represent the
number 3.1415)..
Sometimes one uses $p = \omega + \Omega$ et find
Each of the termes after subtraction
In relation to equinox of reference 1900
One can recalculate i, ω, Ω formulae
Change of equinox classical or matrices
1950 or 2000.
Some orbital elements could be in degres
or in radian. Sometime the polynome result
is big than 360 degres. We must transformed
it to have a result < 360 degres

NUTATION

The axis of the earth undergoes an oscillation
Periodic :the nutation. The position
of the axis without the effect is told mean position
With the nutation position
Axis is called position true.
The moon is first parameter
effect ensues the sun acts. Thus
The nutation is moon-sun
This effect acts on the position of the planetes.
One must thus correct the longitudes and lattitudes.
The quantity $\Delta\phi$ is called correctness
Longitude. La quantity $\Delta\varepsilon$ is called out
Correctness of the inclination of the axis
Formulate giving $\Delta\phi$ +- 0 ".5:
$\Delta\phi$= -17 " .20*sin Ω -1" .32*sin (2*L) -0 " .23*sin (2*L1) +0 " .21*sin (2* Ω)
Formulate giving $\Delta\varepsilon$ +- 0 " .1:
$\Delta\varepsilon$=+9 " .20*cos Ω +0 " .57*cos (2*L) +0 " .10*cos (2*L1) -0" .09*cos (2* Ω)
The values inside of the sinuses and cos
must be in radian. The result $\Delta\phi$ or $\Delta\varepsilon$ has to
be multiplied by:$\Delta\varepsilon$ =$\Delta\varepsilon$ *0.25/900 argument to

have $\Delta\varepsilon$ expressed in degre.

Elements for the computation J2000:

Average longitude of the sun:

L=280.46646+36000.76983*t+0.000 3032*t*t

Longitude average of the Mon:

L1=218.3164477+481267.88123421*t+0.0015786*t*t

Longitude of the rising node:

Ω =125.04452-1934.136 261*t+0.0020708*t*t

The results of L, L1, Ω are in degres

But one must deal result with:

L=L/ 360:L=abs (int (L)- L) :L=L*360

Fill L in for elements

By L1 and Ω and convert in radian dividing

Result by k=180/ pi

EXAMPLE OF COMPUTATION
Δφ Δε

Compute Δφ and Δε For January 1, 1991 at 0 ut
Results:
JD=2448257.5
T= 0.09000684462696783
L=280.1507659162525
L1=100.9130247138484
Ω =299.1300362243322
Δφ=0 ° .004274833929259469
Δε=0 ° .00108289489810969

Compute Δφ and Δε The 16 november 2009 at 6h,18m,14s
JD=2455151.762662037
T=0.09874777993256791
L=235.4625595324112
L1=222.4512570358742
Ω =294.0528758488892
Δφ=0 °.003913379571209027
Δε=0 ° .001004205800128157

There exists formulae with developments series
Incorporating more 100 terms in precision. The add accuracy
Is weak < +-4"in relation to these streamlined formulae.

PRECESSION OF THE EQUINOX

The axis of the earth point verse Northern etoile
And lean between 2 positions maximum and minimal
The earth dish 25800 years to find themselves in same
Position making the turn of the zodiac in the opposite direction
Indeed the motion of the precession of the equinox of the earth
Is retrograde. The earth after one yearly revolution in
The space places themselves a bit in back place
where she was the year before. Each year this motion
Retrograde is measured, with a star's reference
The day reference traditional of the measure is that of the equinox
Springtime called also vernal point.
The earth of year in year draws the long back from the zodiac
Crossing the signs in the sense of the needles
Of one show.
The number is of about 50'' .28 seconds of arc
For period 50 years he is 0 °, 6955565 degre around
For period 100 years he is of 1 ° .3877788 degre
These numbers are data in relation to real measuring
There has millions of years in the past the earth was turning
Much faster on her and the numbers were adding differents.

THE LONG CYCLES PLANETS RESOLVED

The astronomers use equinoxs, epoch of references
To compute the position of the planets:
1900,1950,2000
It allows to be in the real sky because the referentiel is mobile
To change the equinox to find new position of planets, stars, minor planets
several formulae exist. Some use ecliptical coordinates or some
use equatorial one.

INCLÎNATION ΛXIS

The inclination ecliptic is the inclination
of the axis spin of the earth or of the plan
Equator in relation to ecliptic, solar
fundamental plan.
2 inclinations are distinguished:
*Average inclination or « mean obliquity »
*the true inclination « true obliquity ».
The true inclination is the angle equateur
eclipic corrected by the nutation
The other is not.
Formulate skew (inclination:)
ε=23.439 291-0.0130 042*t-0.000 000 16*t*t+0.000 000 504*t*t*t
t=temps in centuries Julian for the epoch J2000.
t=(JDE-2451545)-36525
The axis can seesaw between 2 values
Maximum 24 ° 5 at 24.7 °
Minimal 22 ° 1 at 22 °
Now days it go towards minimun down verse 12300.
For accurate way, the inclination must be corrected

by nutation effect. The moon turn around the earth axis
and have a continuous gravitational effect.
Sometimes the attract is adding inclination axis,
the other the attract is ending inclination axis
Thus the effect is sinusoidal

DISTANCE

The distance R to the sun is expressed
By the formula R=A* (1-e/ k*COS (AE)) ' sun-planet distance
R is called radius vector or radius
That distance is
Expressed in astronomic unity

****Distance to the earth
To compute the distance planet-earth one has
Choice among several formulaes.
I have selected one formula stemming
Nondescript triangles:
a*a=b*b+c*c- 2*b*c*cos (l1-l2)
We have 2 longitudes :
Earth and planet as well as the 2
Distances to the sun and the angle
Make these planets between them

lt=longitude earth
ls=longitude sun
k=conversion en radian

Longitude of the earth

He must add 180 degre to the longitude

Sun

lt=ls+180 :lt=lt/360:lt=abs (int (lt) lt) :lt=lt*360 'longitude earth

dze=abs (lt l*k:)Angle enters the 2 planetes

dist=sqr (r*r+1.000 00011*1.000 00011-2*r*1.000 000 11*cos (dze/ k))

Resultat dist is in astronomic unity

R=distance to the sun

dist=distance earth-planet

METHOD OF THE RECONSTITUTED NODE

Method consists starting with the positions
Recorded by the observatories
to reconstitute the real node which correspond
For that one must go in the
programm and give several values
At the node to find the value fair(α).
Interest of method is than she
Show all of the outsides influences
That undergoes trajectories. More, one can
Identify sequences types. specifics
Planets.
Actually method is a loup, a zoom
Which put in light the slightest inflections
Trajectories.
I chose the ascending node thus in
progress of the computation we have the line
Lo=u+ln1 W).
It means that the node is straightforwardly
Added for result up.
Yet, one can make vary of other parameters

How ω for example or other
The re-enactment ϖ allows to have all
Variations : Longitude of perihelie on one
Hundreds years with the files observatories
We could also appreciate the variations of the distances
And compare them at the used formulae.
These methods of re-enactments of parameters
By report at the real observations are essential.
Because method is using on very long periods of time
And allow to make discoveries than the short

THE MOON

To compute the position of the Moon in geocentric
One must use function as heads inequalities : motion
Lunar develop by E.W.BROWN. Series
have for the longitude 31 terms and 35 for lattitude.

*** computation of position

The results comparison computations
With the recording of the catalogs observatories
Loaded of astrometry: 0.01 ° in RA and declinations

*** method, Measure Computation and Discovery

Starting with RA W50 catalogues
Thus measures of' observatories
I subtract the results of the computation program.
And i get graphic Moon 1964.
The graph show 2 peaks in wide zones
Belonging to the equinoxs
These peaks show an OSCILLATION of the earth.

The continuing measures Moon doing not add
during time. Thus graph shows only
a part of the set oscillation
More certain measures can be filtered,
suppressed because they do not fill
The criterion accuracy demand argument for
Catalogue
Thus if one wants study this oscillation
Quantitative and qualitative, some
One must not leak measures during time,
In the zone of the equinoxs and elsewhere.
Why?
Because the moon is a fast planet
Who by consequent is going to develop
The slightest oscillations earth
Indeed in addition to the zones equinoxs
The earth has to have impulses
Gravitational irregular and
random rebounds who
Are able to demonstrate themselves at all moments
This methods differences for the Moon
And method of the node reconstituted argument for
Mercury and Venus can realize
The oscillations earthly with means
Of astrometries.
Yet, Graphs realized could induce gravitational
Discovery.

MOON 2

Some of the computations of the Moon with coefficients e.w. BROWN

aml=mean anomaly moon

ams=mean anomaly sun

p=mean elongation moon

q=mean distance moon from ascending node)

8735 LO=lo+769.02 *SIN (2*AML)

8740 LO=LO+36.12*SIN (3*AML)

8746 LO=LO+1.94*sin (4*aml)

8750 LO=LO+0.11*sin (5*aml)

8755 LO=LO+0.01*sin (6*aml)

8760 lo=lo+2369.9*SIN (2*P)

8764 LO=LO-668.11*SIN (AMS)

8768 LO=LO-125.15*SIN (P)

8772 LO=LO+4586.43*SIN (2*P AML)

8776 LO=LO-411.61*SIN (2*Q)

8780 LO=LO+211.66*SIN (2*P-2*AML)

8785 LO=LO+205.96*SIN (2*P AML AMS)

8790 LO=LO+191.95*SIN (2*P+AML)

8795 LO=LO+165.15*SIN (2*P AMS)

8800 LO=LO+147.69*SIN (AML AMS)

8805 LO=LO-109.67*SIN (AML+AMS)
8810 LO=LO+55.17*SIN (2*P-2*Q)
8815 LO=LO-45.1*SIN (2*Q+AML)
8820 LO=LO-39.53*SIN (2*Q AML)
8825 LO=LO+38.43*SIN (4*P AML)
8830 LO=LO+30.77*SIN (4*P-2*AML)
8815 LO=LO-45.1*SIN (2*Q+AML)
8835 LO=lo-28.47*SIN (2*P AML+AMS)
8840 LO=LO-24.42*SIN (2*P+AMS)
8845 LO=LO-18.61*SIN (p AML)
8855 LO=LO+14.58*SIN (2*P+AML AMS)
8860 LO=LO+14.39*SIN (2*P+2*AML)
8864 LO=LO+13.9*SIN (4*P)
8850 LO=LO+18.02*SIN (P+AMS)
8872 lo=lo-9.37*sin (2*p-aml+2*q)
8874 lo2=sgn (lo) :lo=abs (lo)
8879 LO=lo*0.25/900:lo=lo/360:lo=abs (int (lo) lo) :lo=lo*360
8868 LO=LO+13.19*SIN (2*P-3*AML)
8912 lo=lo+lml*k:lo=lo/ 360:lo=abs (int (lo) lo) :lo=lo*360
8914 lo2=sgn(lo):lo=abs(lo)
8879 LO=lo*0.25/900:lo=lo/360:lo=abs (int (lo) lo) :lo=lo*360
8880 lo=lo*lo2 'lo=longitude geocentric in degree

8915 la=18461.48*sin (q) -6.3*sin (3*q) +0.005*sin (5*q)
8916 la=la+1010.18*sin (aml+q) +999.69*sin (aml q) +61.91*sin (2*aml+q)
8930 la=la+623.66*sin (2*p q) -6.49*sin (ams+q) -4.86*sin (ams q) -5.36*sin
        (p+q) -4.79*sin (p q)

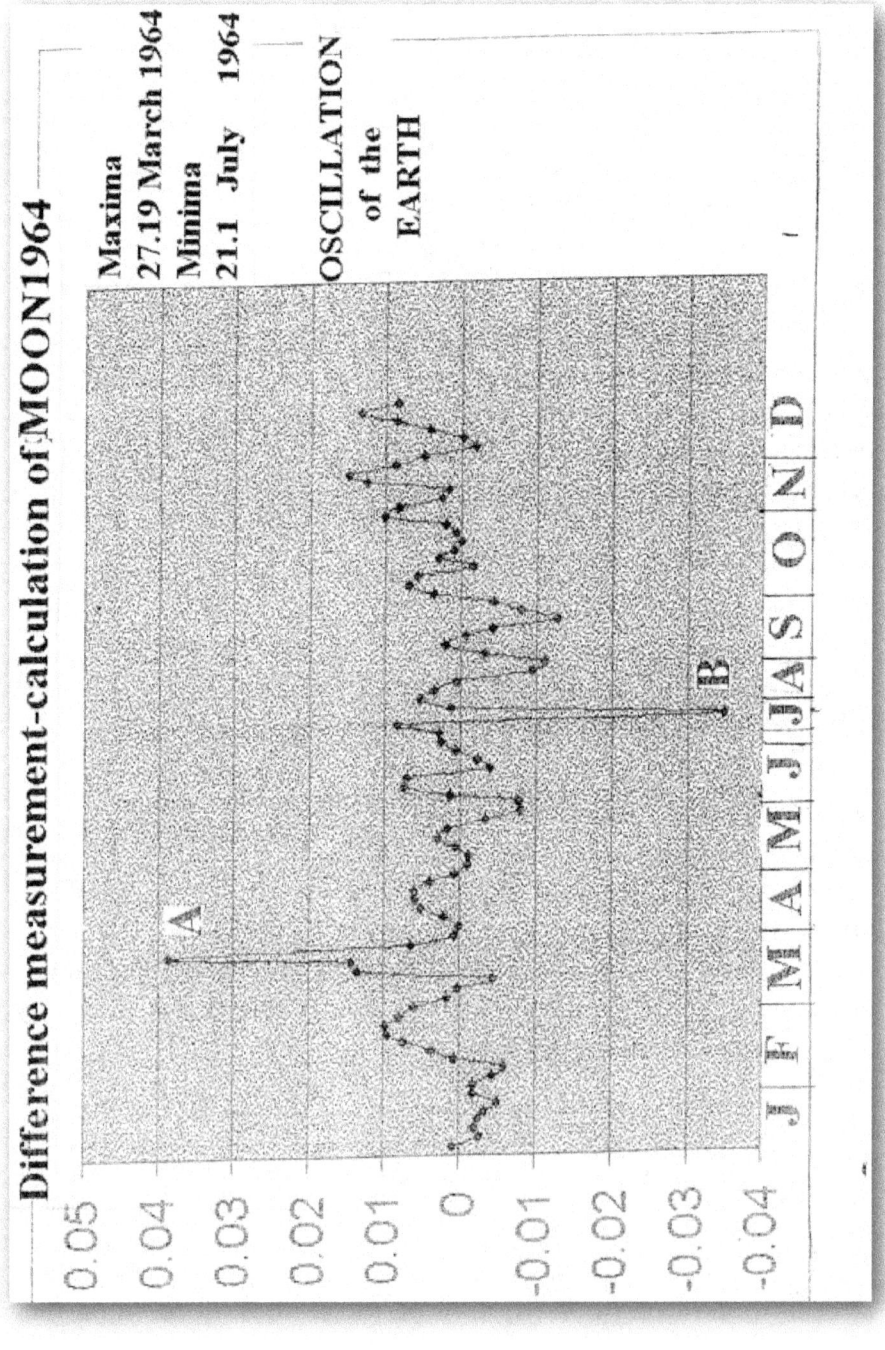

MERCURY

First planet of solar system having for peculiarity
An inclination more high I +- 7 as well as of some excurtions
Maximum mattering.
She is wholly mapping by spatial probes
With radars analyses surface

*** computation of position

The comparisons with the differents catalogs show:
W50 1963-1971 on the totality period measures
Differences RA like 0.01° to 0.011°
And 0.01 ° for the declinations.
Series 1964 -1965 (Tokyo) has differences
lower 0.01° in RA and declinations.
W1 1977-1982 differences under at 0.01°
In RA and declinations.

*** reconstituted node and Discovery

Starting with the positions RA 1964 of the catalog W50
I get the graph with the correspondent shape
Mercury 1964 (see graph.).
I notice presence of 1 peak of important amplitude
concerning the period mi- May to mi- June.
The origin of this oscillation is:
One OSCILLATION EARTH
Gotten with means of astrometry
This oscillation of the earth happens at some
Places specifics of its trajectory.
The zones where these oscillations are acting
Correspond at the moment where the earth pass
of declinations positive to negative and conversely :
The equinoxs March and September.
Afterwards with measures catalogue W1 1978
I get Mercury 1978 bends her of the reconstituted node.
In 1978 the oscillation is acting mi- May to mi- April
The oscillation is more complete thus after the peak
Negatif a less important peak in amplitude is noticed :
the peak positif. The oscillation seems unbalanced
Amplitude because the measures are discontinuous.
He lacks some days in measures.
After comparison I notice oscillations
Similar on the bends Moon and Venus

*** complete Enregistrement of the oscillation
So that the oscillation be completely reproduce

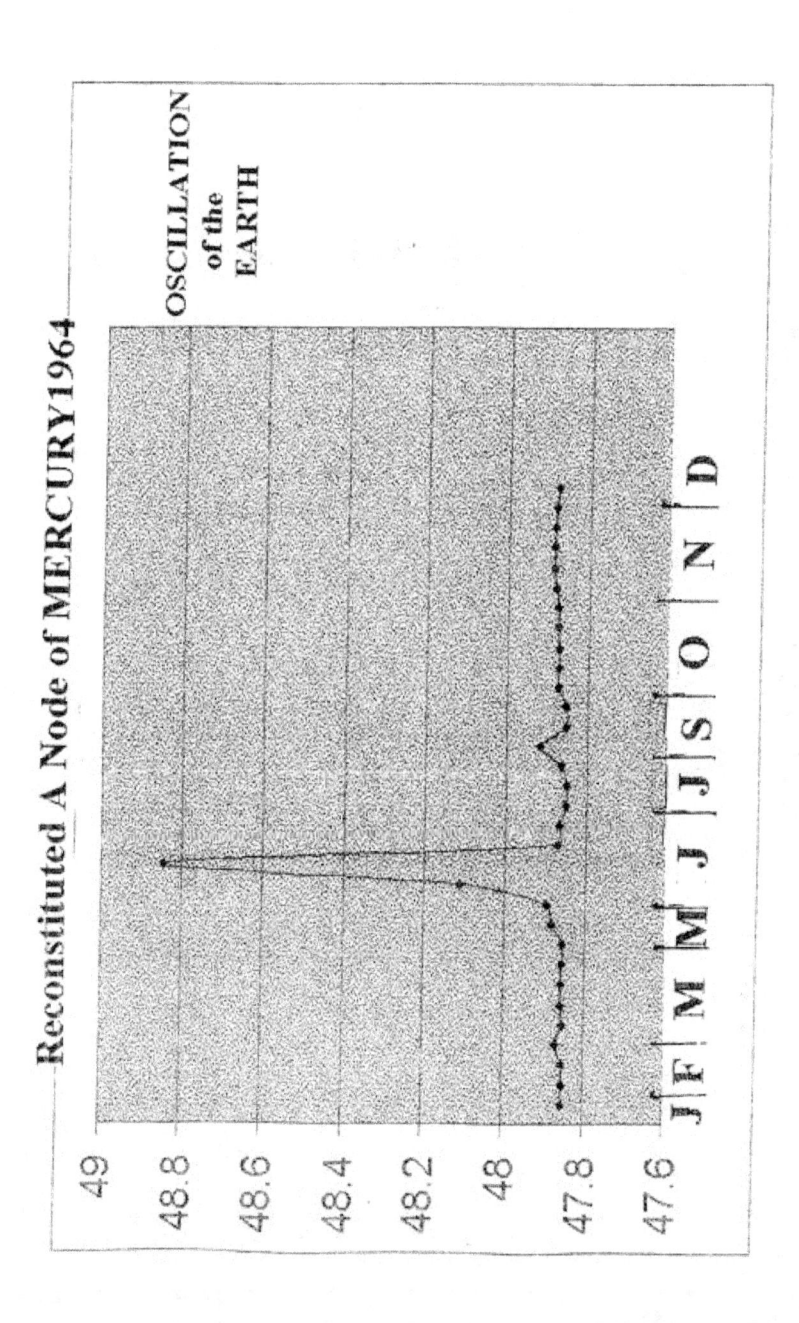

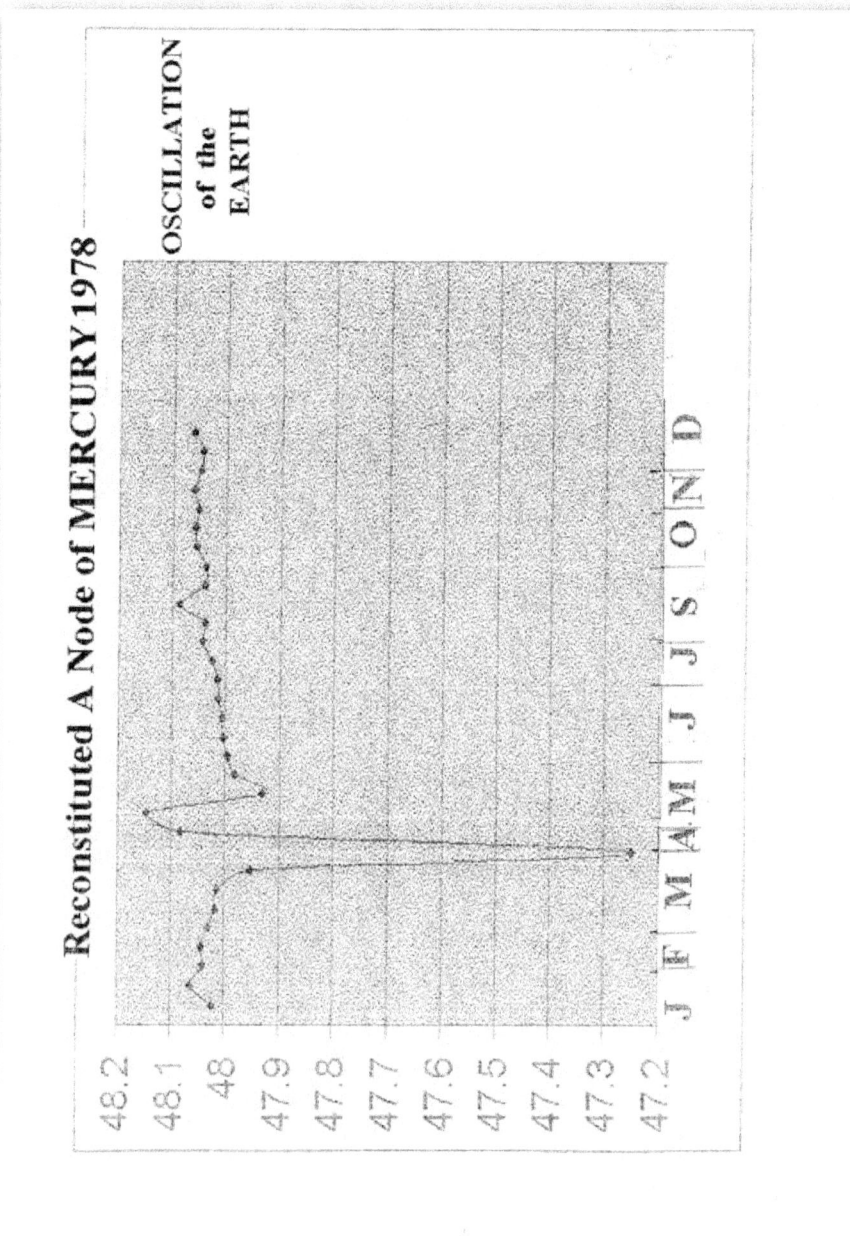

VENUS

Second planet solar system, she belongs
inside. Known since the antiquity this
planet have like caracteristic one
excentricity weakling goes e=0.006
theirs elements make her a regular planet.
Her atmosphere is cloudy, acid and hot.
Its surface has been mapping by radars
Boarded.

*** computation of position
The comparisons with the catalogs show:
W50 1963-1971 gives differences
Of 0.01 ° in RA and declination.
W1 1977-1982 gives differences
lower 0.01 ° in RA and declination.

***Reconstituted node and Discovery

Starting with the positions RA of the catalog W50
Year 1964 i get the graph of the node

Reconstituted Venus 1964.
On this graph the presence is noticed
2 ripples in the zones of the equinoxs
March and September.
In fact, this is two OSCILLATIONS of the earthly globe
A first oscillation of weak amplitude A
is acting during the period mi- March to mi- May
In zone equinox March.
Ensues one second oscillation B amplitude
Mattering is acting during mi- aout to mi- September
In zone equinox September
The second oscillation have one
Smaller peak who is a peak of amortization
Of oscillation. Next planet pick up again
Normal path.
One can also note as Venus seesaw
Weakly hanging its trajectory

                                                ***

Also the re-enactment of the node
With the catalog W1 1978 show
a part only of the oscillation.
Why? Because
Many measures carry out have been filtered,
cut thus her do not fill in
The criterion of accuracy Catalog.
One must thus in the equinoxs zones
Conserve the wholly set of the measures
Carried out day by day (streaming).

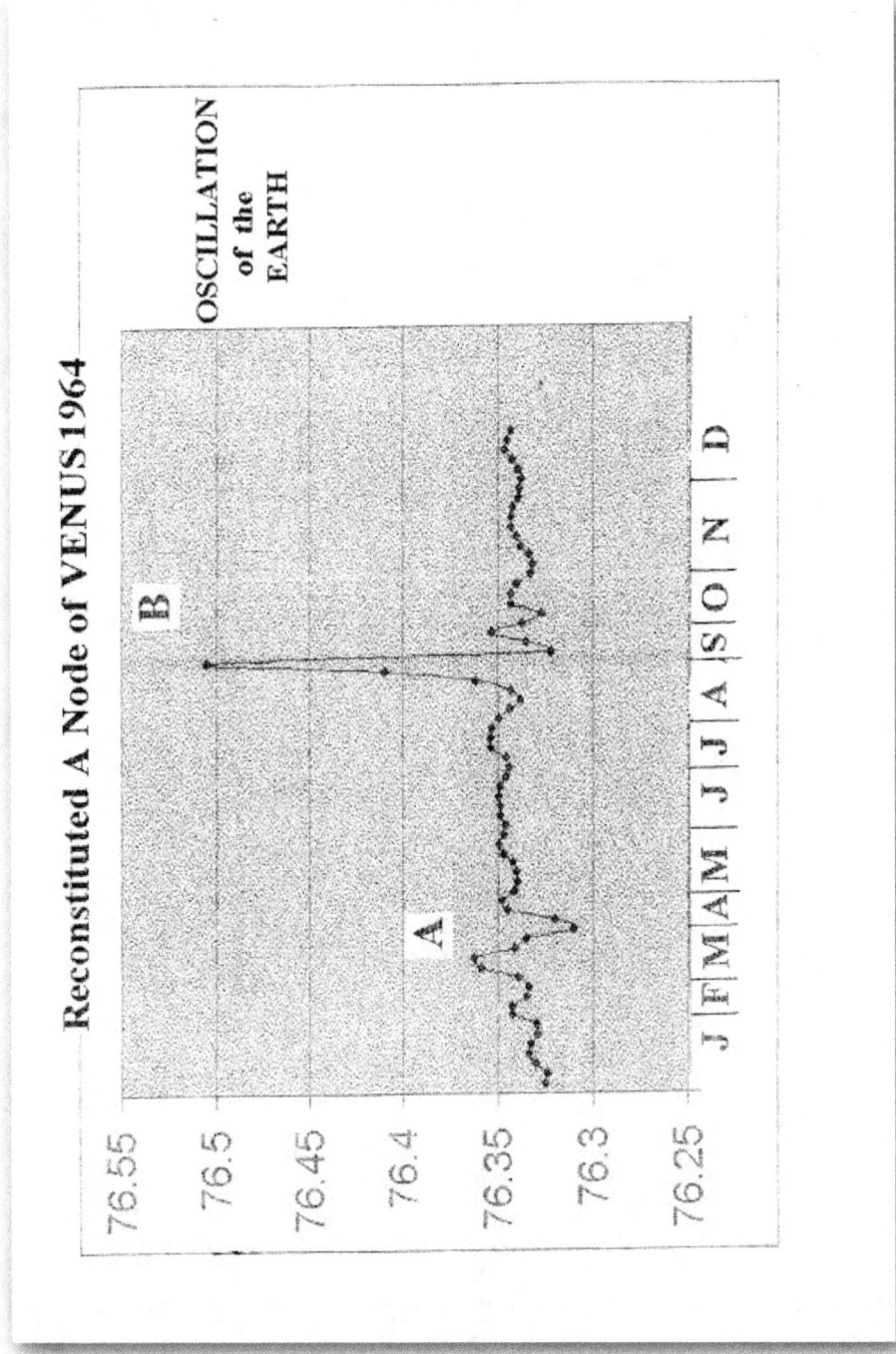

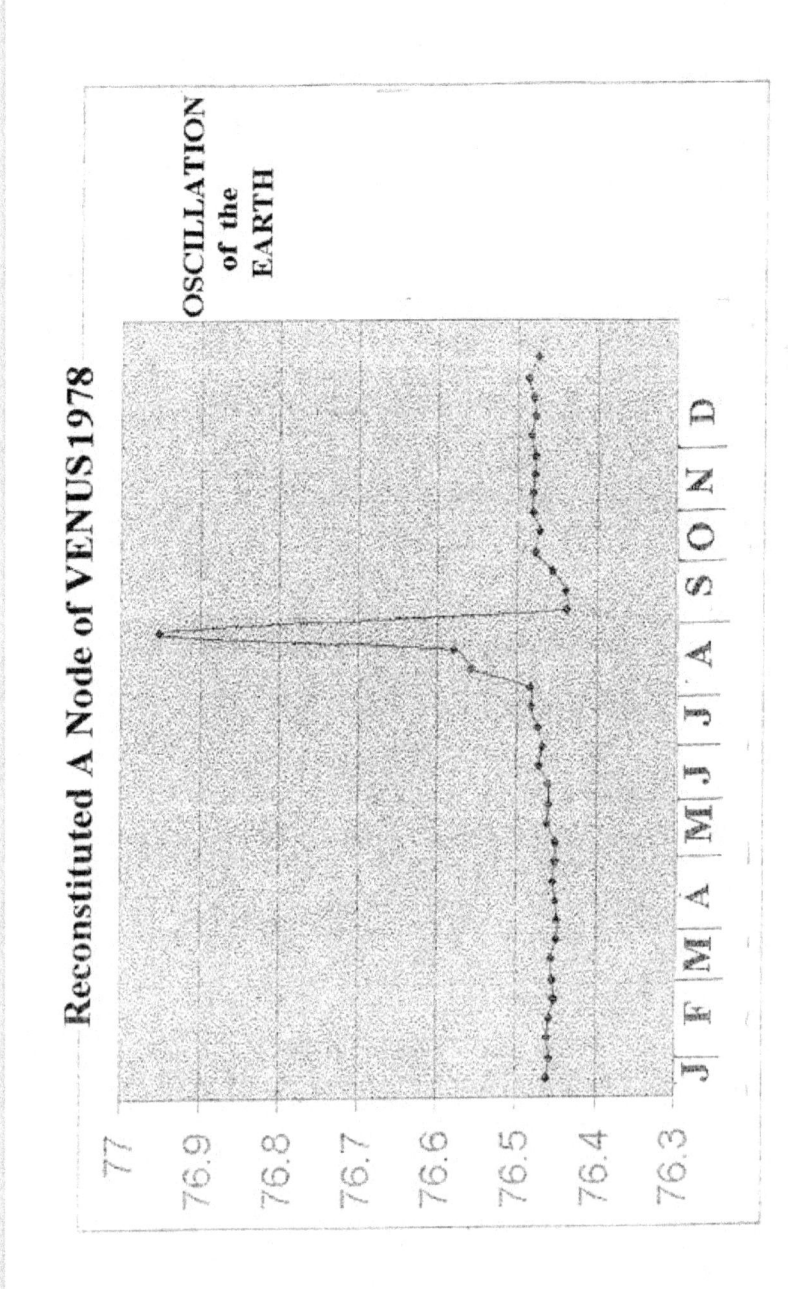

MARS

First of outers planets, Mars is the object
Of very numerous studies. The surface is mapping
with radars aboard in spaceship.
Analyses spectrometrics complete
The device of studies. Its place themselves
on the probe or at the ground (vehicle robot)
These studies have arguments for organize
The colonisation of planet by the man.
The planet is comparable in grounded size
And have weak luminosity.
Sometimes because of elliptic trajectory
Mars approach near the earth (1 ua)

***Comparison of its position with
The catalogs:
*Series tokyo (1949-1962)
These series vary little in right ascension
In relation to the computation 0.025 to 0.01 argument for
Date the most recent.
*W50 1963-1971

same variation is noted
In the RA 0.018 to 0.01
*W1 1977-1982
Variation of the resultts
From .0.02 to 0.005
*W2 1986-1995
Variation from 0.01 to 0.035
*Carlsberg series (May 1986 at November 1986)
Variation 0.01 at 0.035
The set of the results of year different or identical
show variations which are incorporated in
The interval 0.01 at +0.03. These variations are always
In exces, every catalog in relation to the computation is.
** one takes the series W2 and Carlsberg of 1986
Identical differences are noticed by report
At the computation retorted like this :
January to July 0.01 verse 0.04 degres
July to december 0.04 verse 0.01 degres
Besides I noted a distorsion specific
With the computations when Mars comes
At the most near the earth. Its
Disappear when Mars move away.
This report is true whatever
Nature of the turn near the earth:
Wide turn or tight.
In fact the first part of the turn
(the coming) is more inaccurate than the computations.
the second part is correct

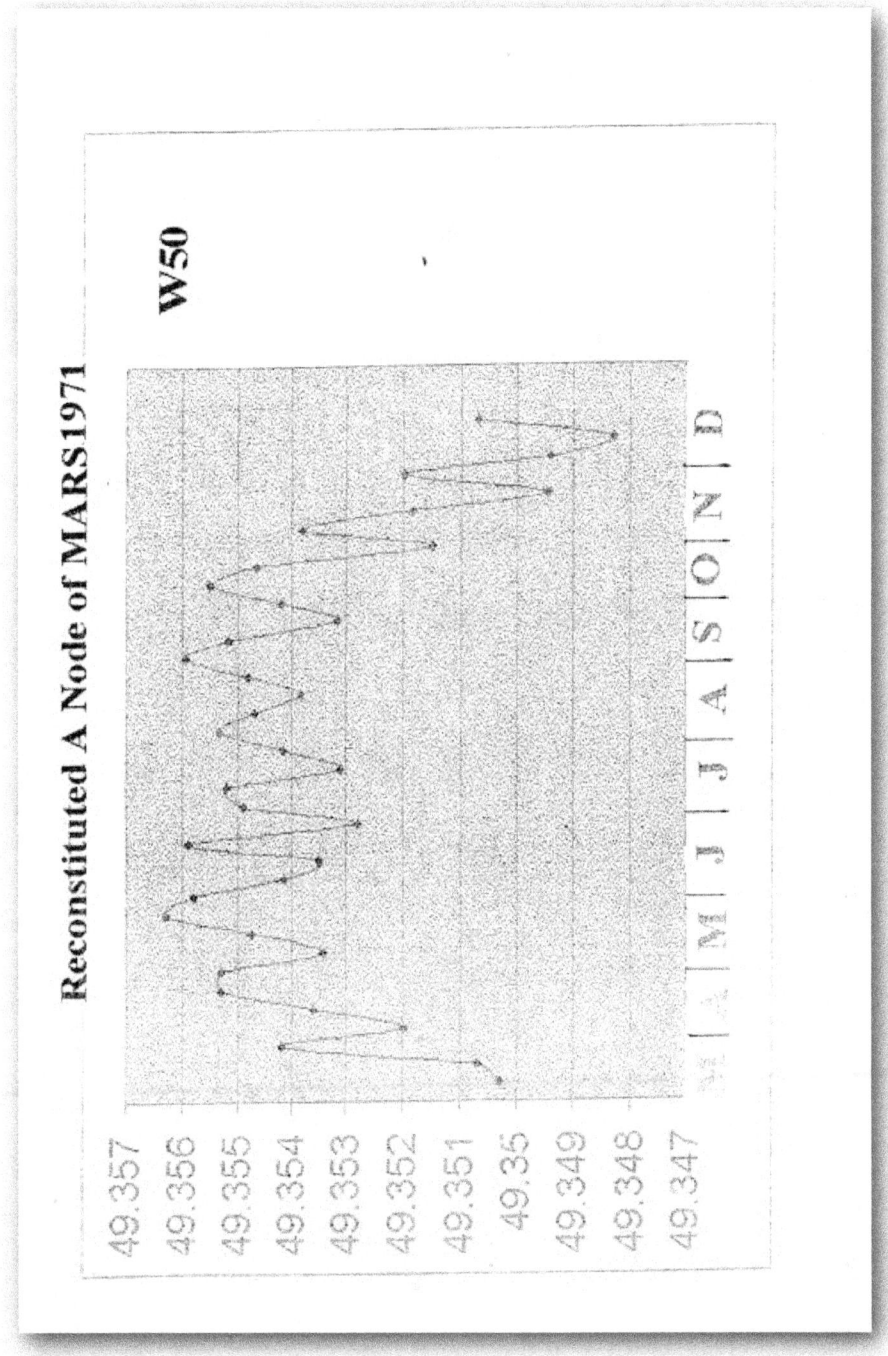

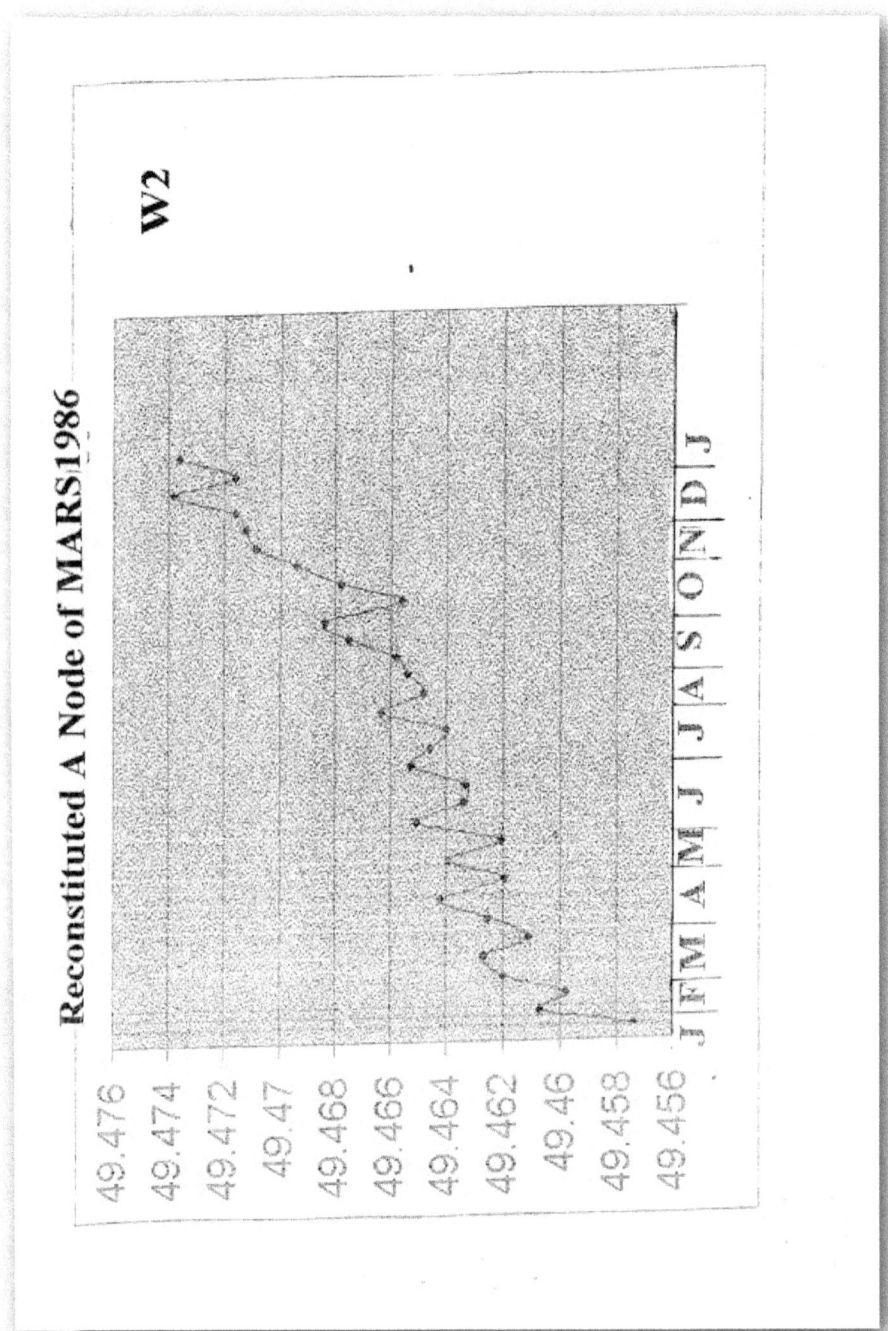

JUPITER

One of bigger two planet system
Solar, She belongs to the group of giants gaseous
that 's the most near the sun.
She have numerous satellites naturals

****calcul of position

With the computation of position of Jupiter
We reach the problem of correctnesses
Trajectory.
**the theory the most used is the one
gravitational interaction with
The others giants outers planets of which
Head is Saturn.
With the measures carry out by
Observatories from 1850 one can
easely verify all types of influences.
If one compares the conjunctions Jupiter Saturn
And the measures carried out by the observatories :
Trajectories are noticed each

2 big planets are not modified
While they are the most near, the attaction
Each of them is maximum thus the effect depend
Of the square of the distance F= G*m1*m2/d*d2.
More the planet is far, more the attraction is weak and conversely
As mathematics formulates describe the attract.
*More one can verify that there is not
Correlations with alternations of the reconstituted shape
With the conjunctions Jupiter/ Saturn or Jupiter/ planets.
Conjunctions Jupiter/ Saturn:
4 december 1901
September 23, 1921
24 august 1940 to the 23 february 1941
22 february 1961
8 february 1981 to 7 august 1981
These conjunctions are seen of the earth and with the effect
Retro- gradations.
**La second theory take notice
Barycentre Sun planets
With integration numerical of the trajectories
Of the set planets.
Following the variations of barycentre
Correctnesses are applied
At the trajectories.

How get a forecast accurate
Trajectory with the right ascension ?
***Dans this goal i use the reconstruction
Bend of the rising node with
The positions measured of the observatories
With the relation LO=U+LN1 (ln1= Ω)
I made vary LN1 (rising node)

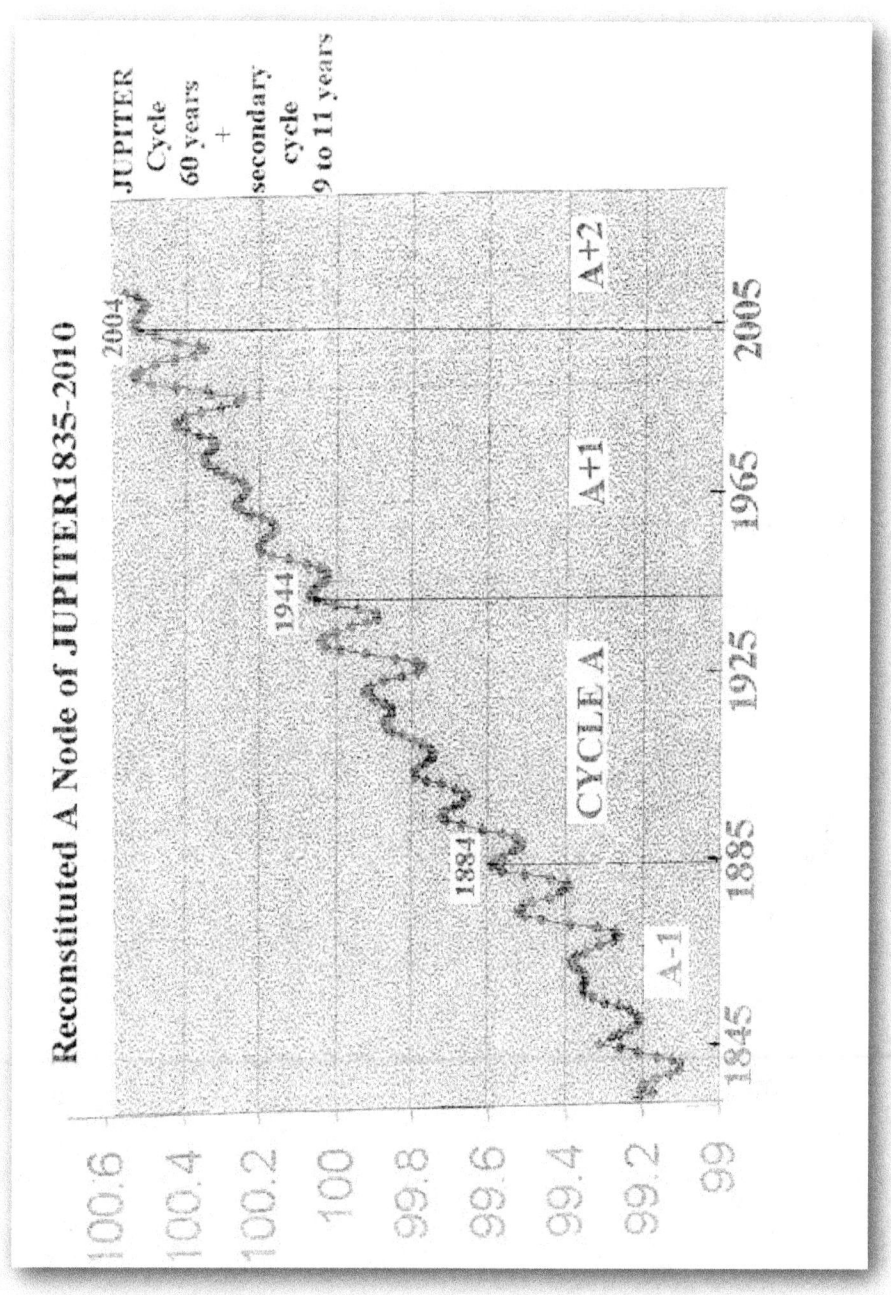

Reconstituted A Node of JUPITER 1835–2010

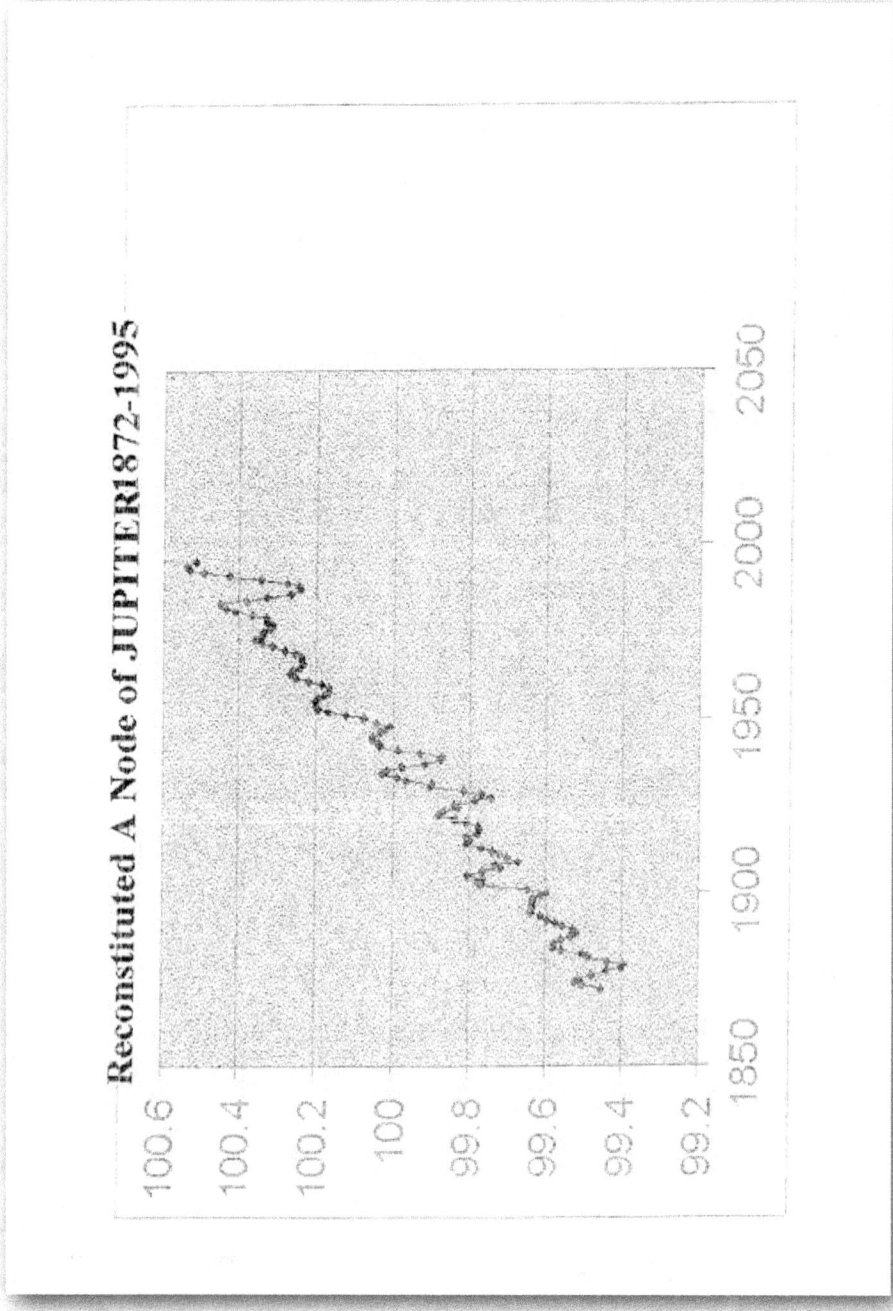

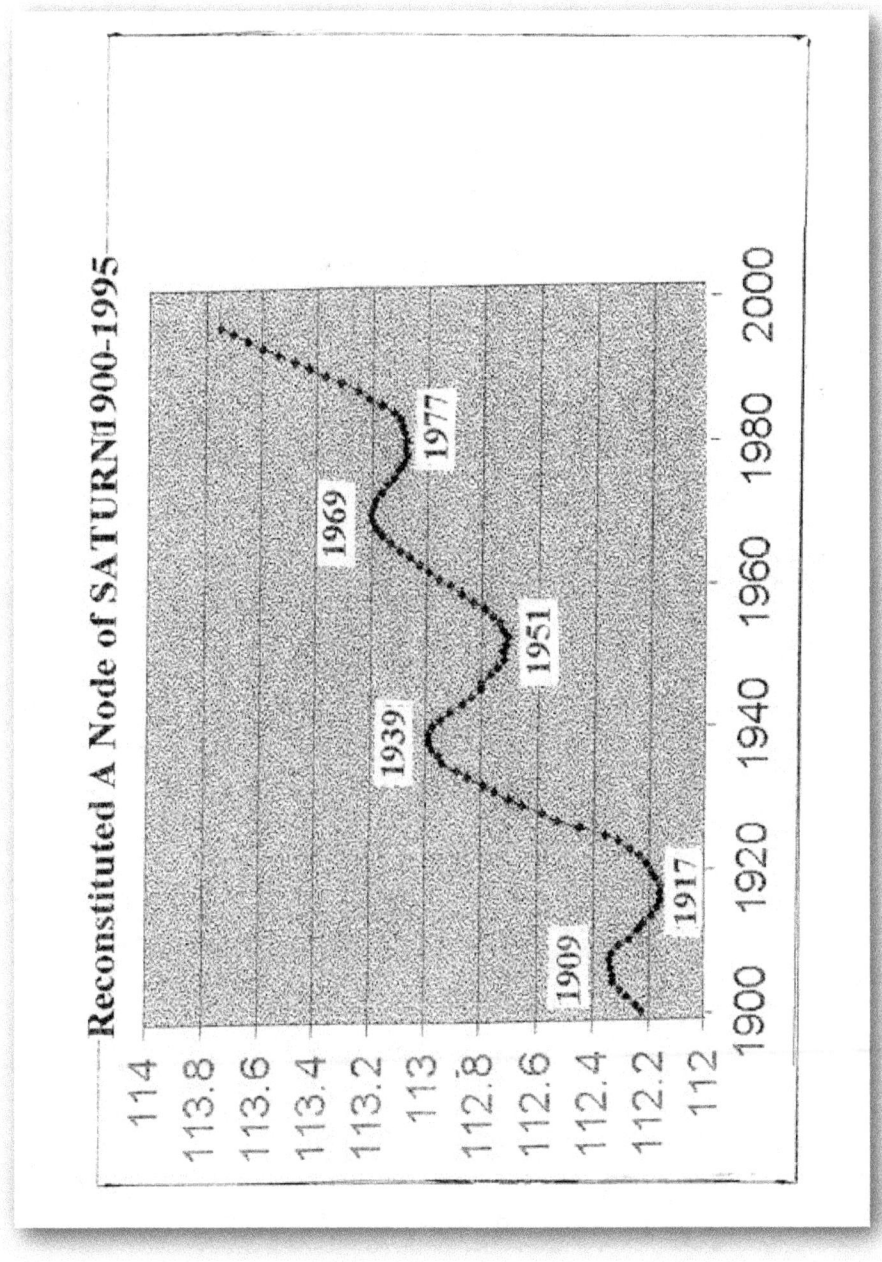

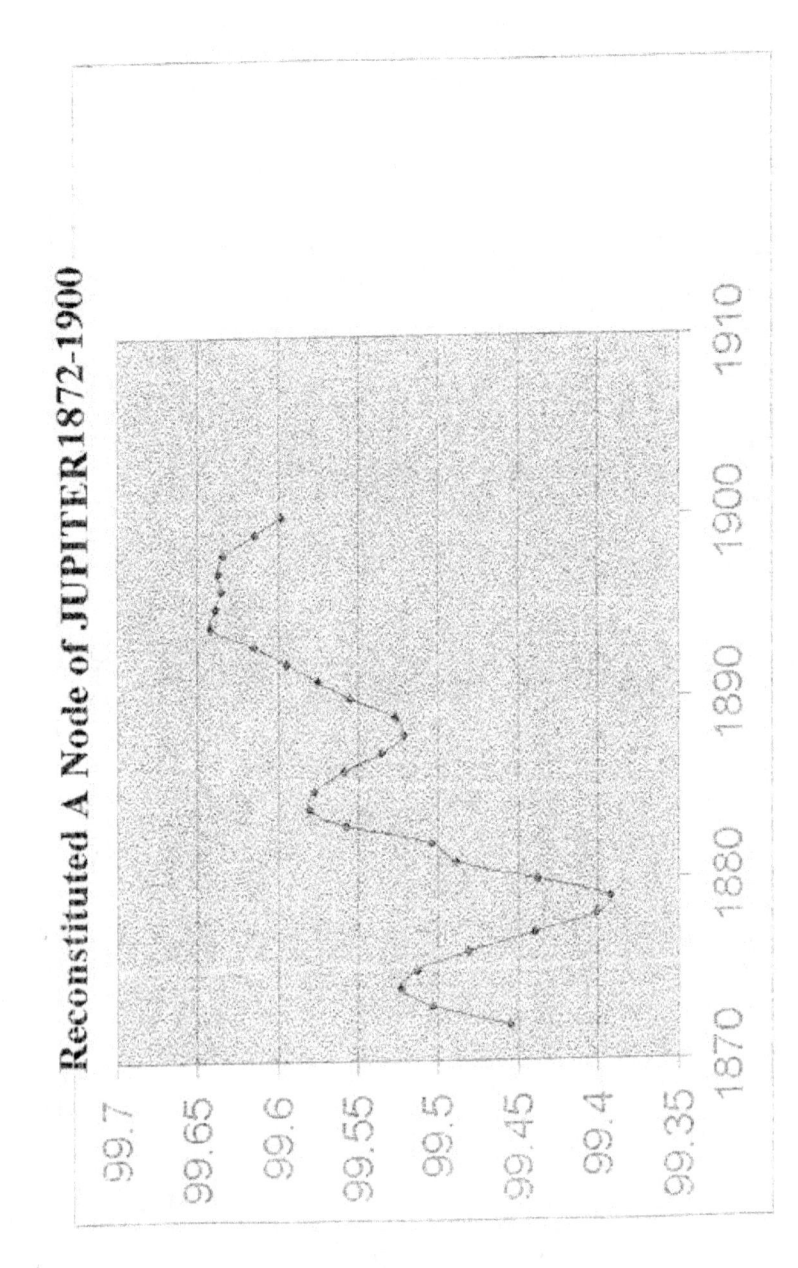

Reconstituted A Node of JUPITER 1872-1900

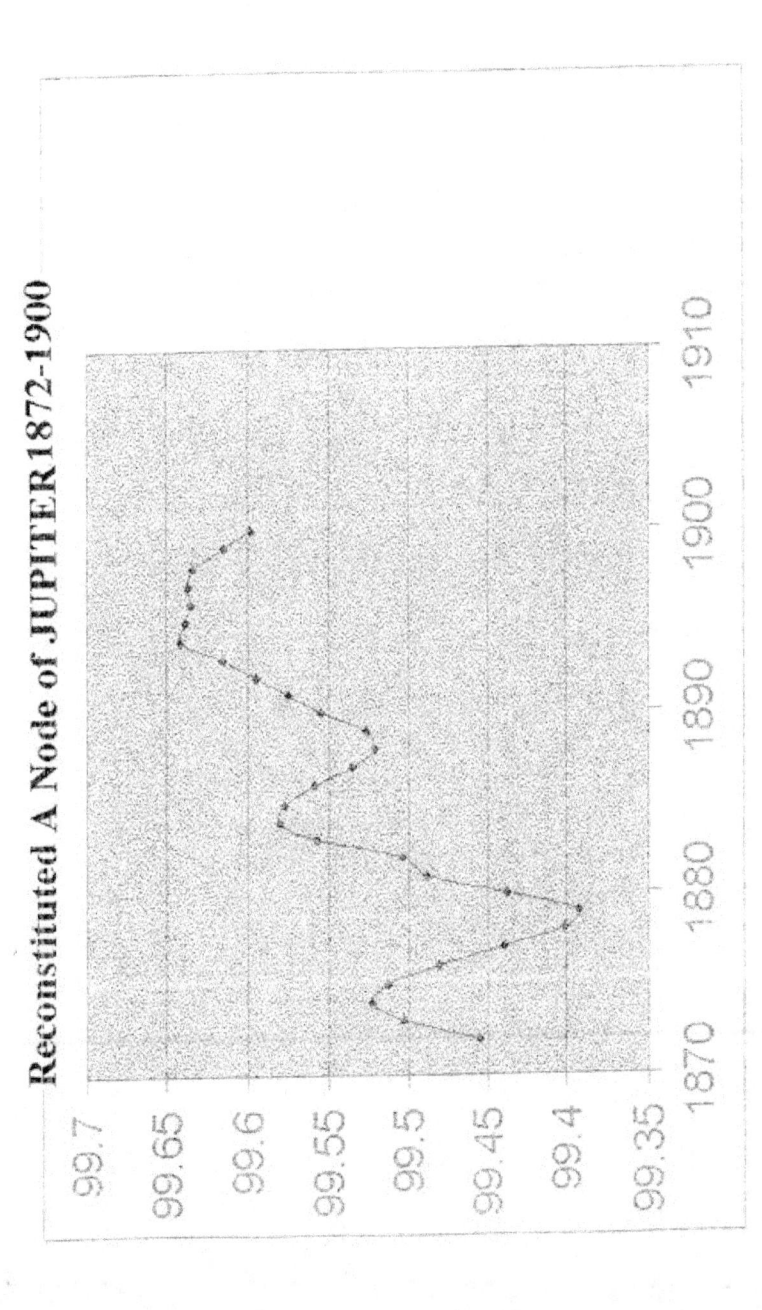

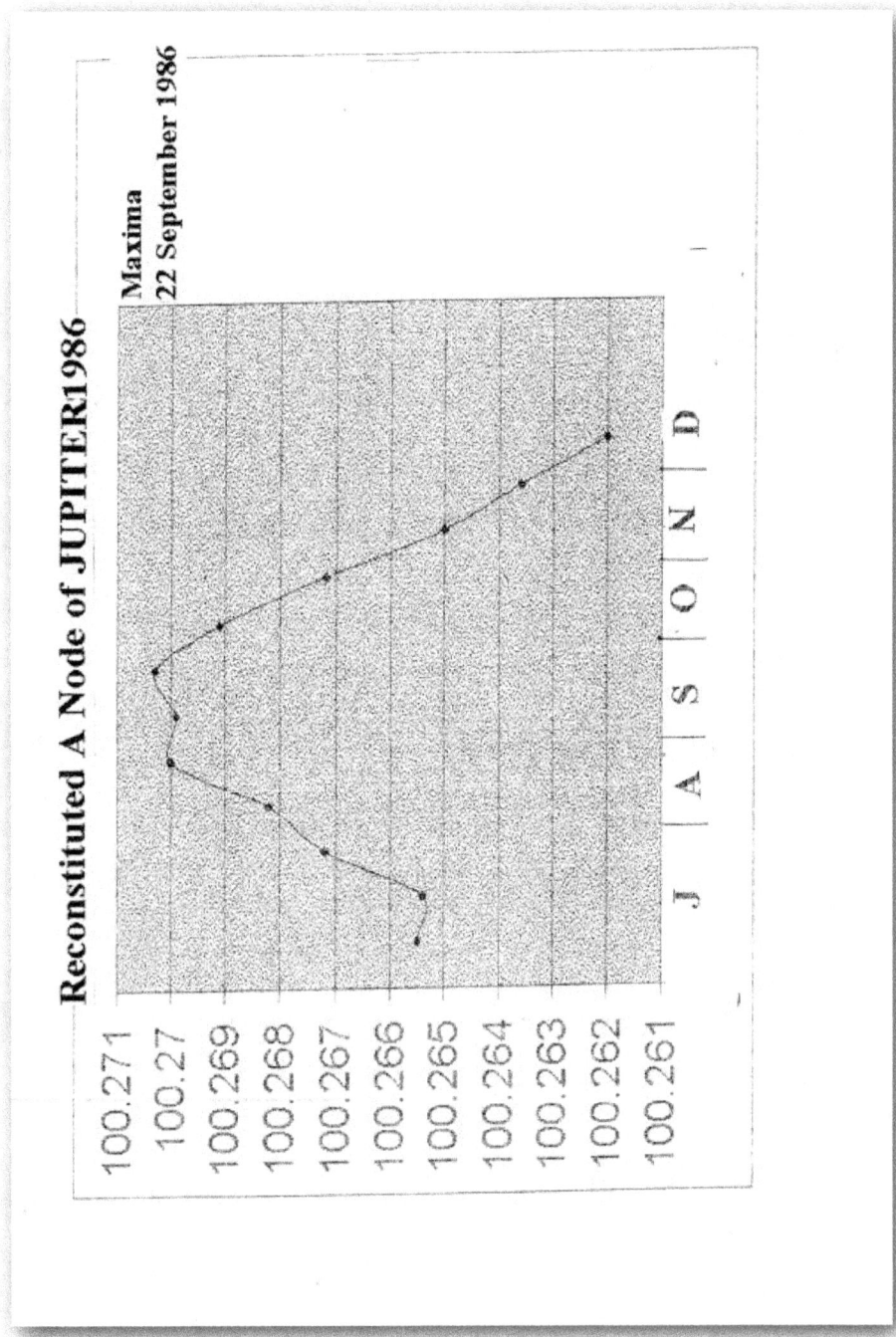

SATURN

Second more big planets system
Solar, Saturn is famous for these rings.
she have numerous satellites natural
**Calcul of position
The problems of correctness trajectory
Have a long time problematic on the plan
Mathematical development.
Correctnesses of more 80 terms
In longitudes have been tried, .Formulae.
For the means terms arrive around 0.2
Degre of difference (RA) with the observatories.
For the long term one arrives easily
At 0.4 or 0.5 (RA). Only the professionals announce 0.01
degre with formulations more complex than the plain
orbitals elements.
If we compare forecast positions of several professionals
after 20 years the difference is more than 1 degrés
and augment with time

Fortunately recordings
Observatories from 1830 and before
have clarified these singularities
Trajectory.
***Reconstituted node Rising

With the relation LO=U+LN1 (ln1= Ω)
I made vary LN1 (rising node)
To get the right ascension gotten
By the observatories.
I get the reconstituted bend of 1834-2001
Of Saturn (graphic see) This bend shows alternations
Minima and maxima. The bend must have of the shape ax
linear right.
All deformations of the bend point out a distorting force
External at the one of the trajectory planet.
This bend present one repetitive regular cycle composed by a big
Peak sinusoidal and a second littler in amplitude. The wholly frequency
of this cycle is 60ans= 34. ans +26 years. This cycle of 60 years
That i chose arbitrarily start in 1879
Is repetitive and superimposable. Indeed the cycle 1879-1939
Is superimposable by translation graphic on 1939-1999
This motive themselves repeat constant and regular.
Which is the cause of these ripples?
Cause of these ripples is interferences between the inertia
of the big mass, the fast and the gravitation.
This modulated gravity push and attract planet the long its
Trajectory progression. After the maxima the node progress
Perpendicular to the axis towards the minima. The high
Peaks and the second are built same manner.
Bend is meaningful of SATURNE, its speed and
its mass and its distance from sun. En fact this is resultant gravitational
between planet and Sun.

***kinetic

The planete execute 1 tower around the sun at an average weak distance
And she ensues execute one second at a average distance more big by
Effect of compensation kinetic (forces of recall)
In relation to the others outers planets
One can note the absence of secondary cycles
It will be obligatory to expect carried out series with
Satellites of observations (Hubble, Hipparcos)
To know if secondary cycles exist
In the basic cycle Saturn

***analyse of the signals

If one assimilate the bend of the node to a trajectory
Electronics it is noticed that she is constituted
Triangular signals deforméd by the speed of planet.
This one composes itself of a big triangle and a small

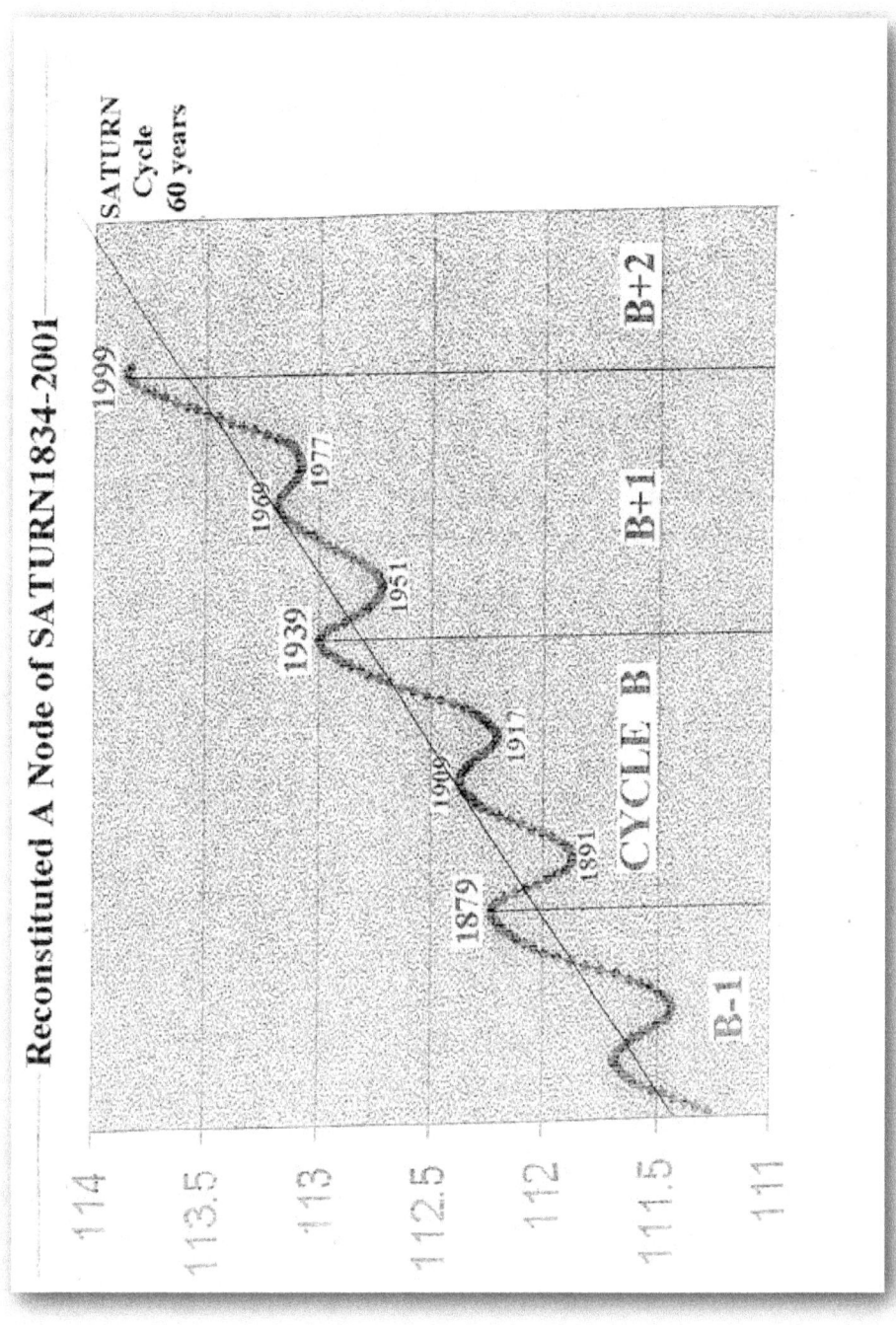

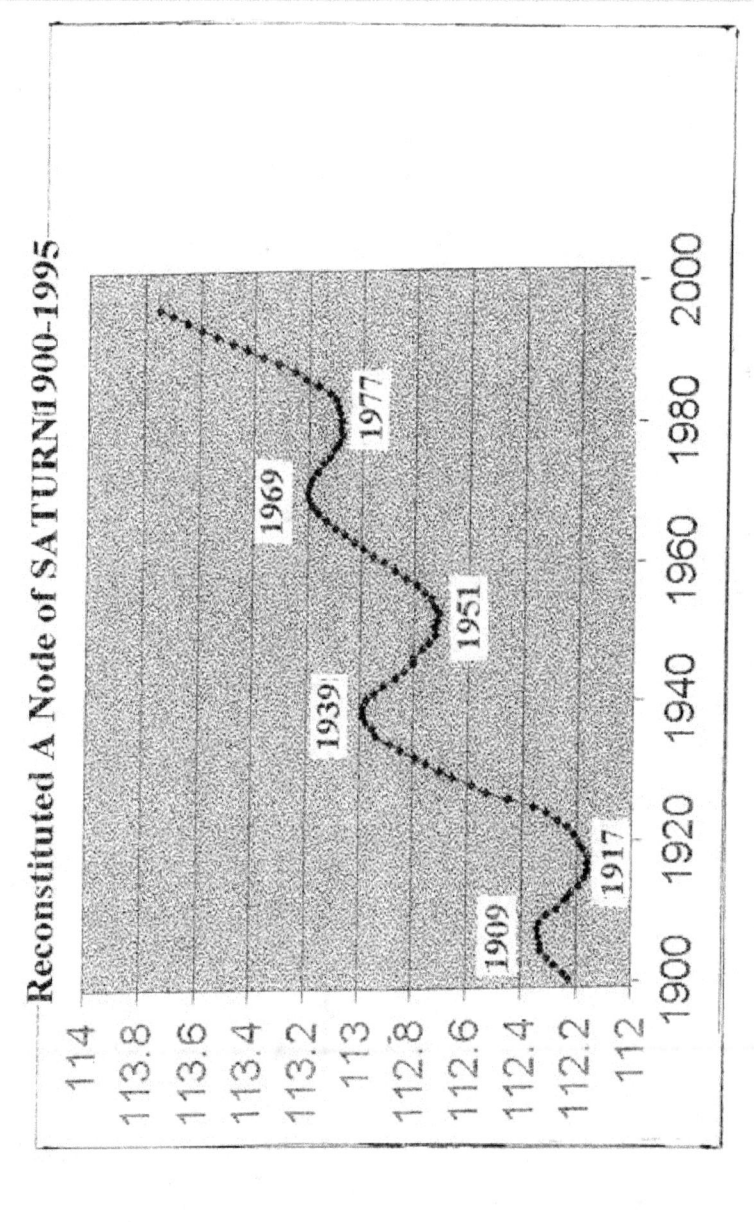

Reconstituted A Node of SATURN 1900-1995

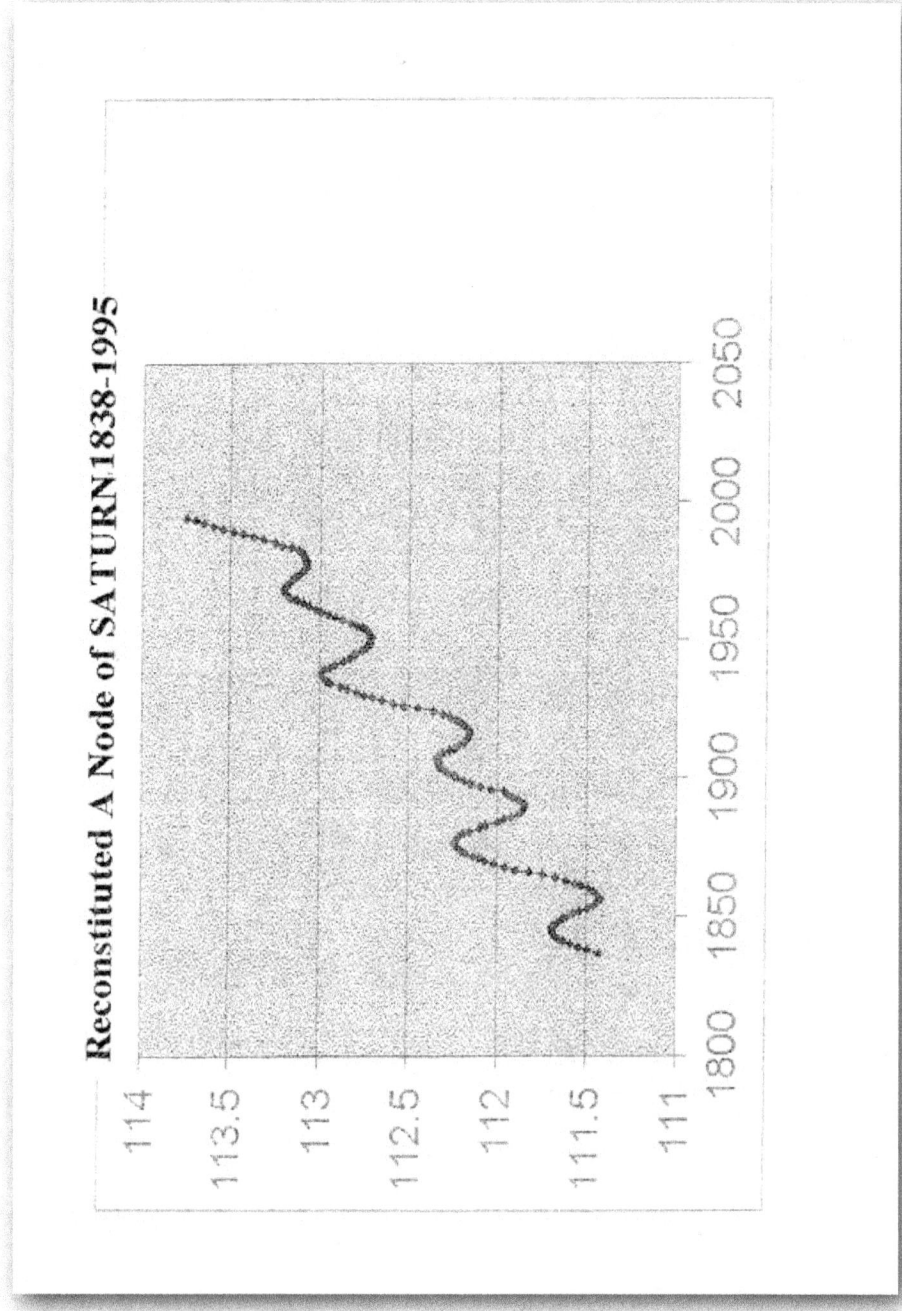

Reconstituted A Node of SATURN 1838-1995

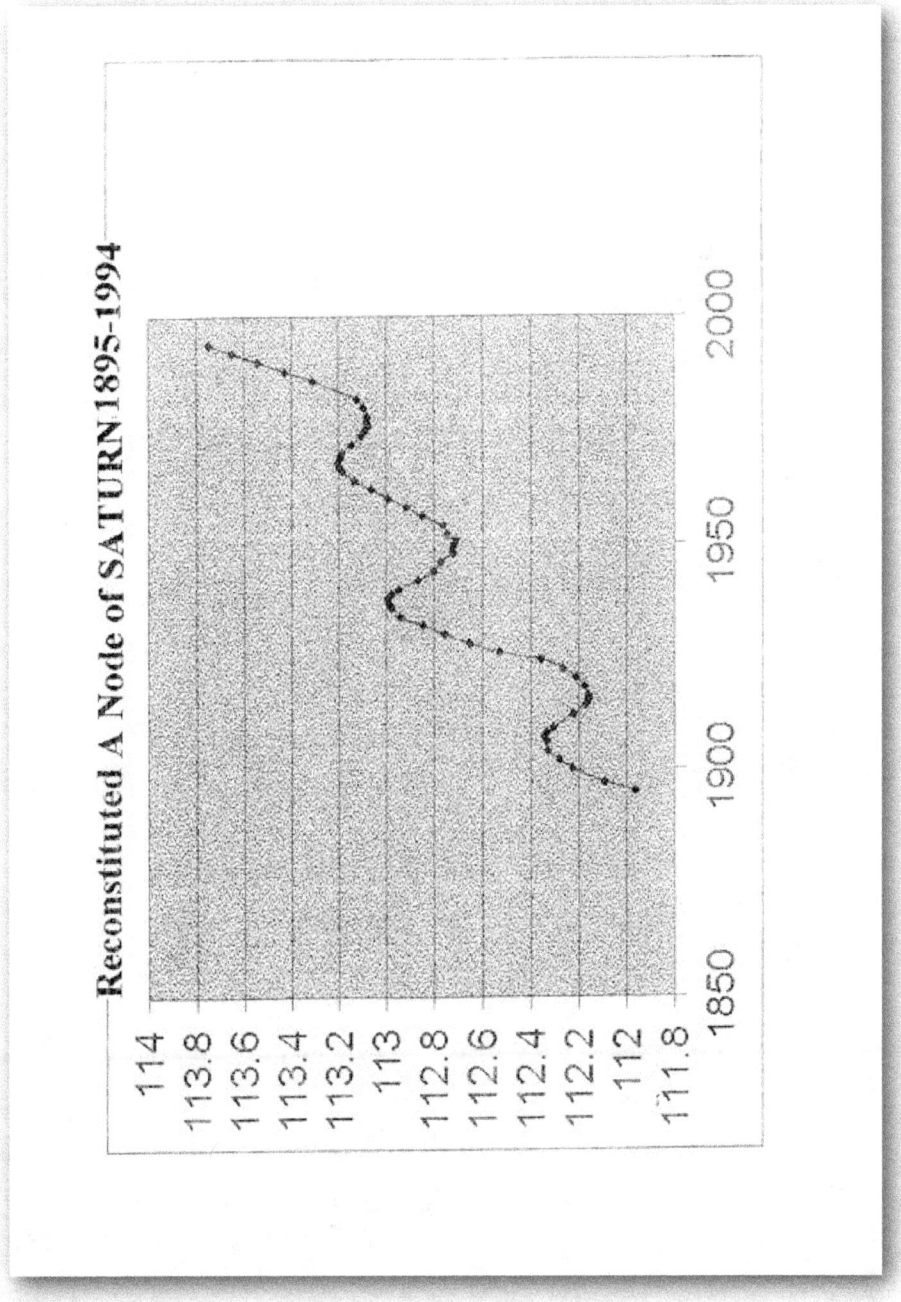

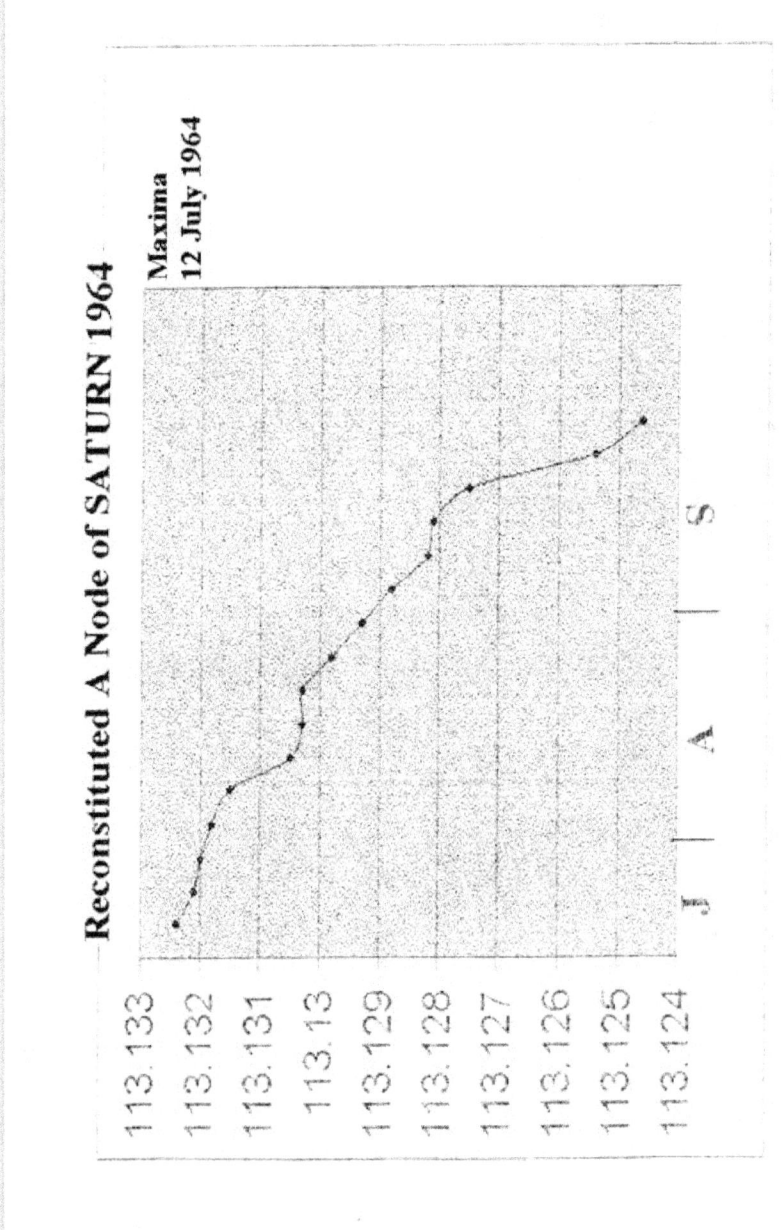

Reconstituted A Node of SATURN 1964

Maxima
12 July 1964

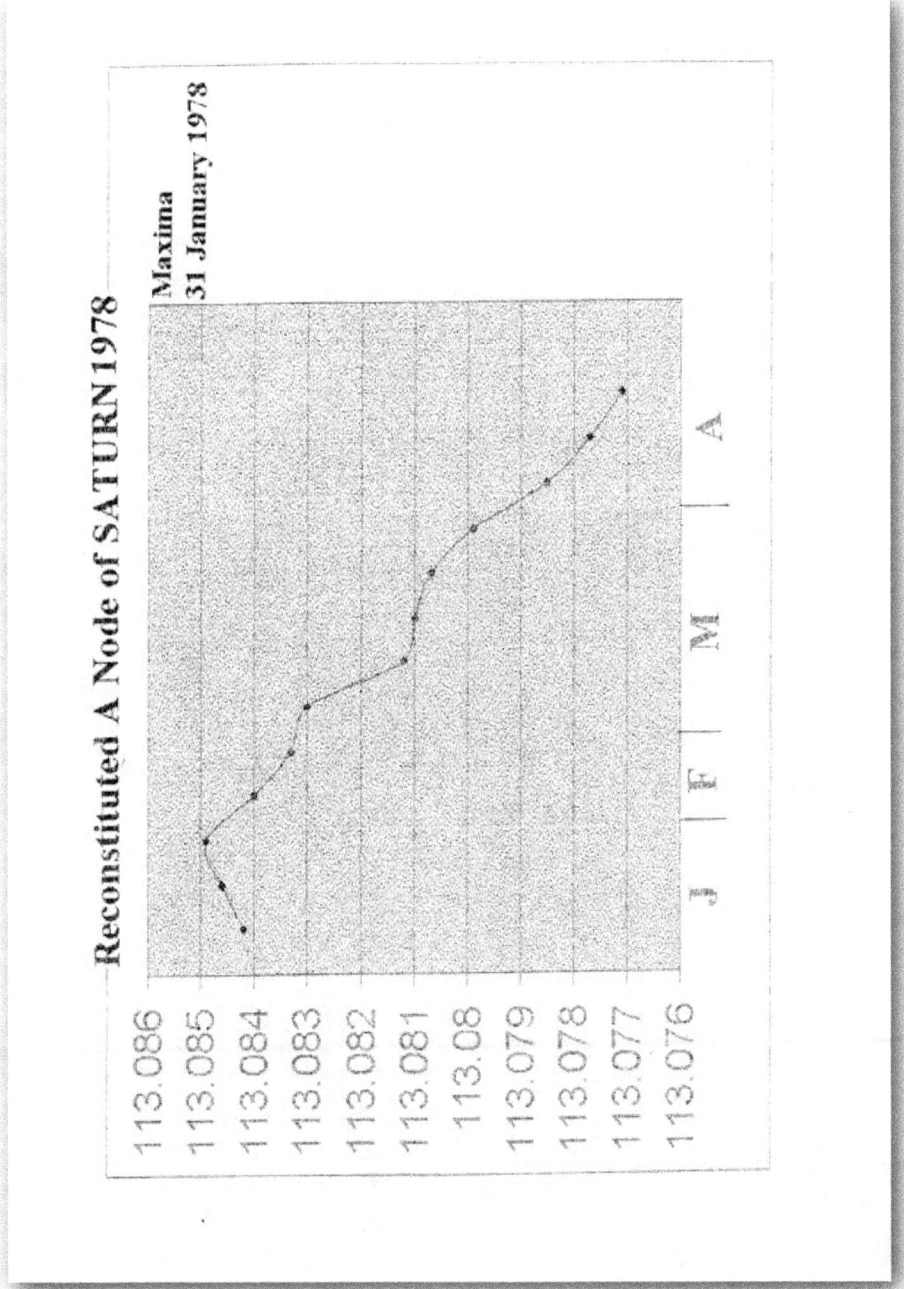

Reconstituted A Node of SATURN 1978

Maxima
31 January 1978

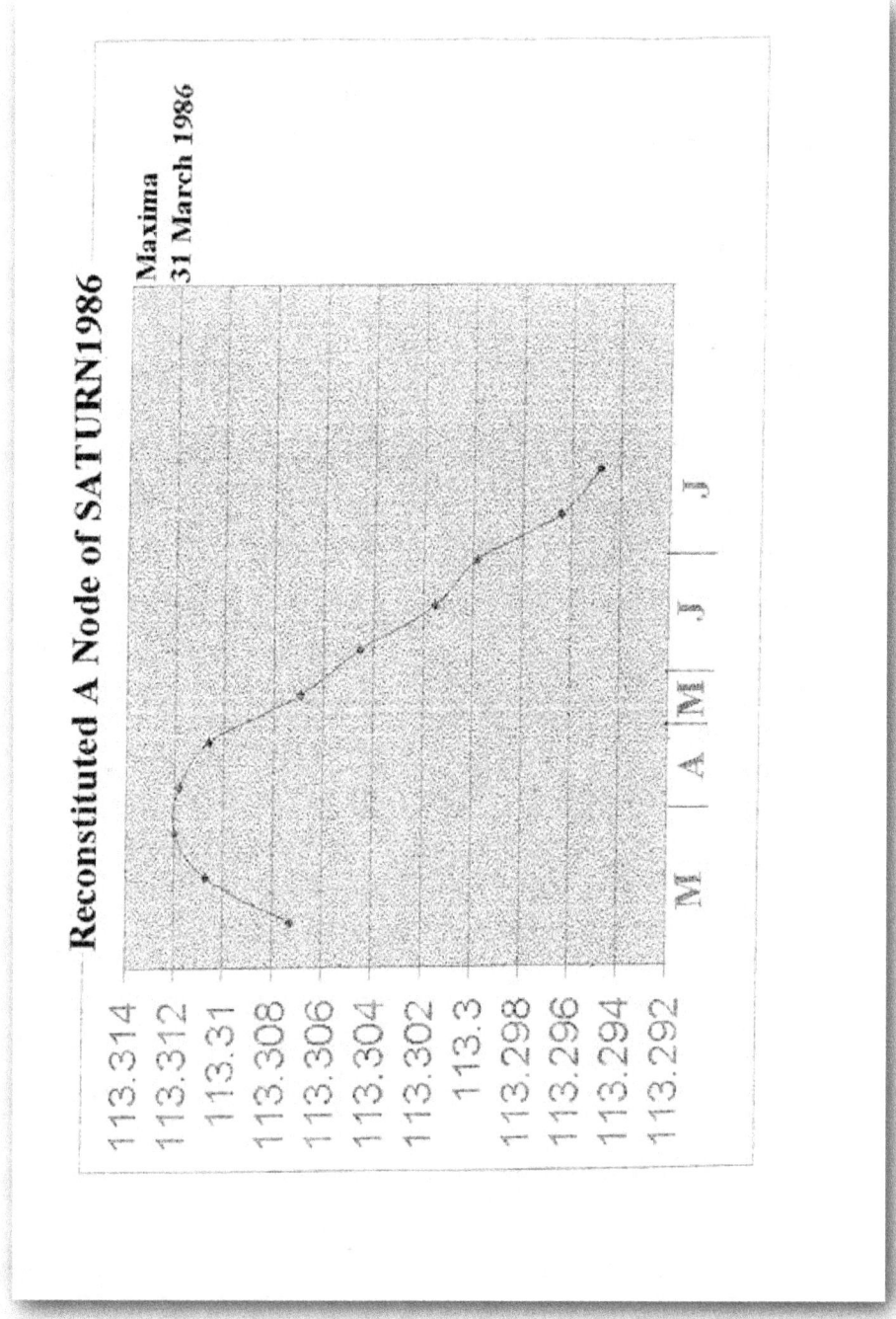

Reconstituted A Node of SATURN 1986

Maxima
31 March 1986

URANUS

Georgium Sidus, first name of this
Planet become uranus. Giant gaseous
Her have rings and satellites natural

*Calcul of position
Theory the most used is the one of an interaction
neptune with accurate computations
On long periodes (correctnesses secular) or short.
I cite the conjontions uranus/ neptune
31 december 1820 to September 9, 1821
25 february 1992 to November 25, 1992
To show that the conjonctions don't deform bends
Rising node reconstituted.
To know the trajectory of these planet
We have the recordings of the observatories
From 1800. More we have recordings
Recent of the sattelite Hipparcos. These placed satellites
In the ionosphere between 600 km and 800 km provide

accurates observations.
The recordings who allowed to build
The reconstituted bends are:
1801-2008 USNO U.S.A.

***Shape of the rising node reconstituted

For this aim i use relation $LO=U+LN1(ln1=\Omega)$
And i was making vary LN1 (rising node)
With the positions observed
The curves are reconstructed from 1800 to 2050.
With the main scheme one can observe the cycle
From Uranus which repeats itself every 84 years.
Indeed, the shape of a ring may be superimposed
A than any other. This cycle breaks down into 2 parts:
A long part is the descent towards the sun (1925-1975 = 50 years).
After the perihelion the planet moves away from the sun
More undulating (1975-2009 = 34 years) to the aphelion
* Ripples.

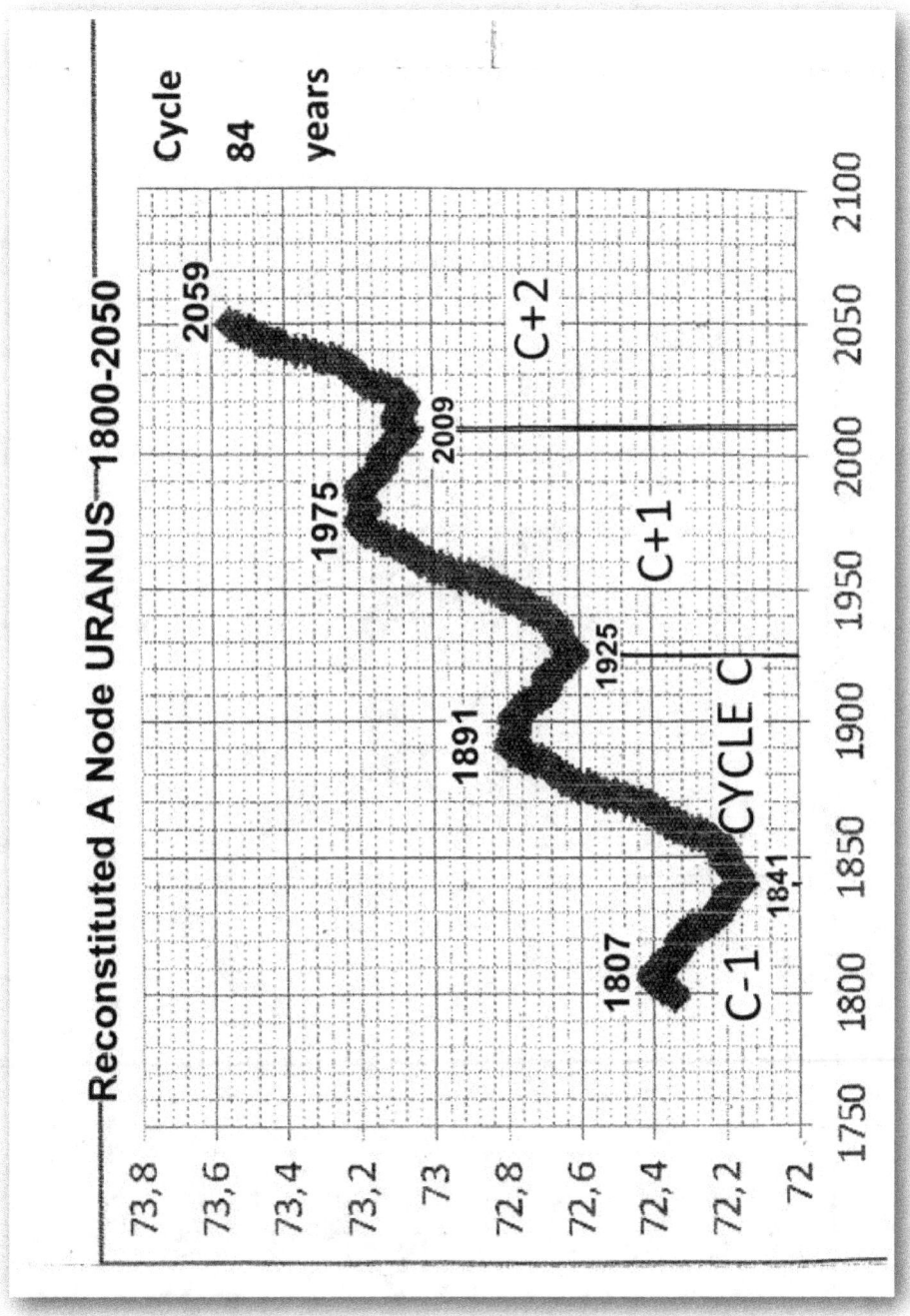

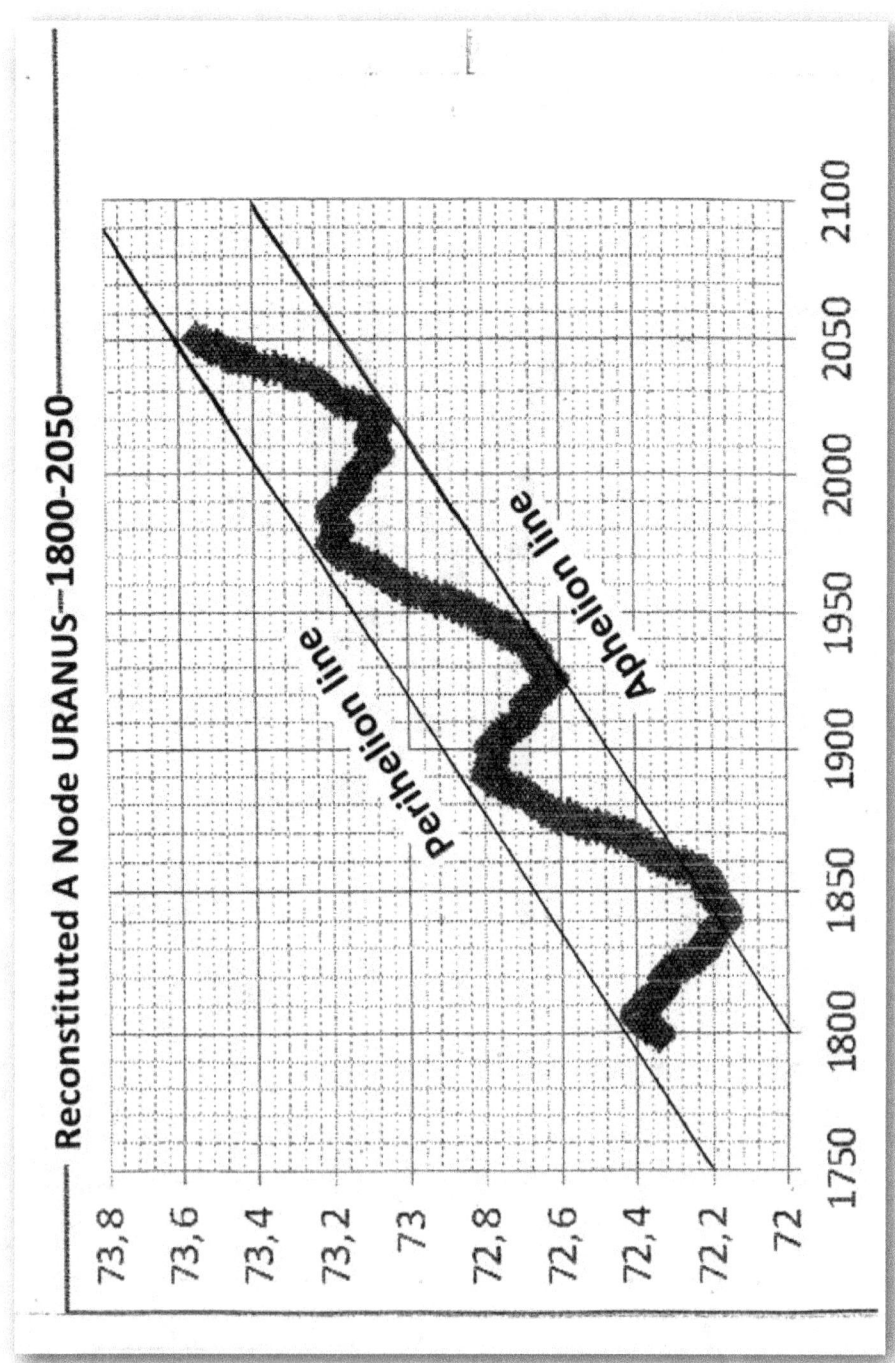

Reconstituted A Node URANUS—1800-2050

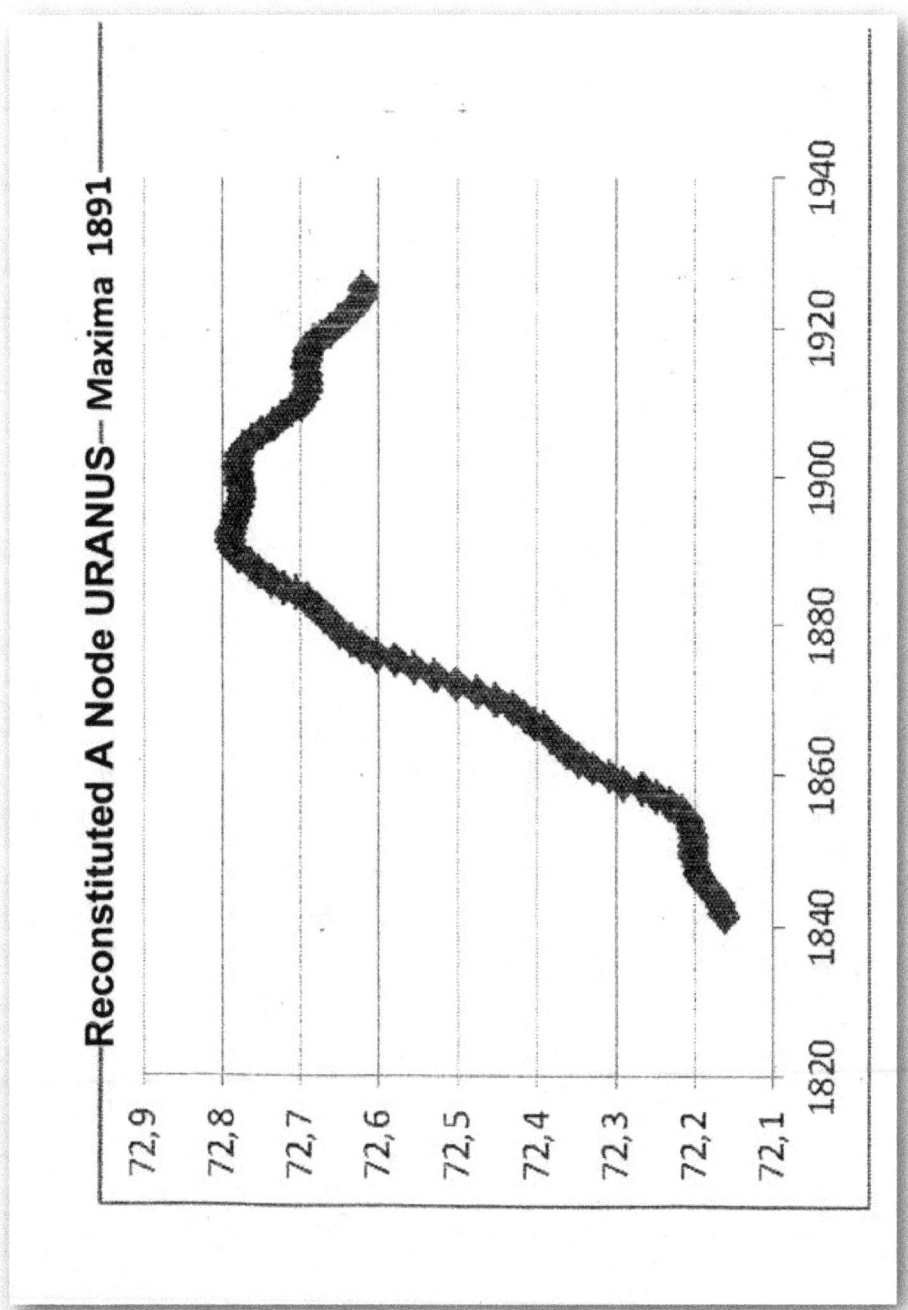

Reconstituted A Node URANUS— Maxima 1891

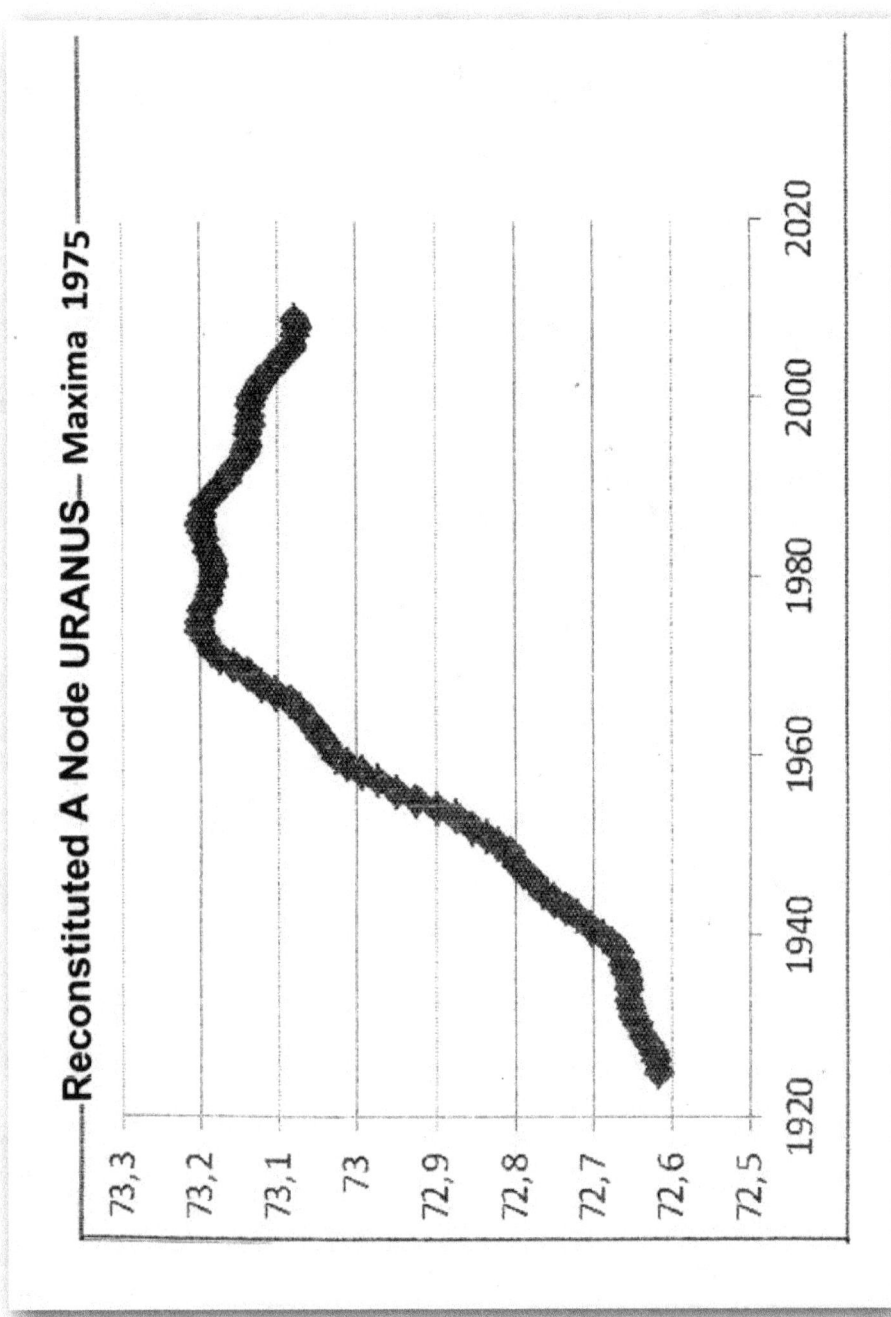

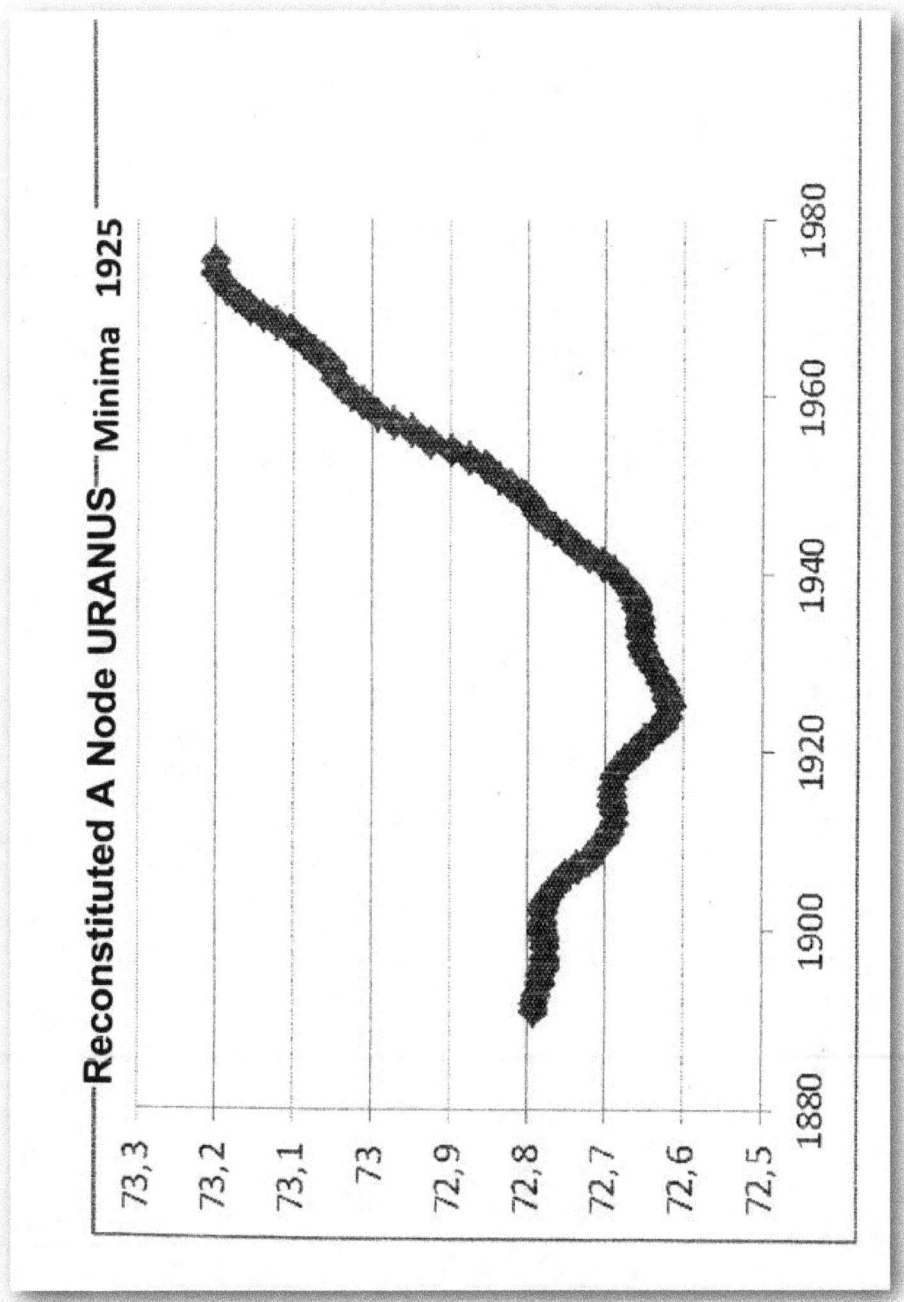

Reconstituted A Node URANUS—Minima 1925

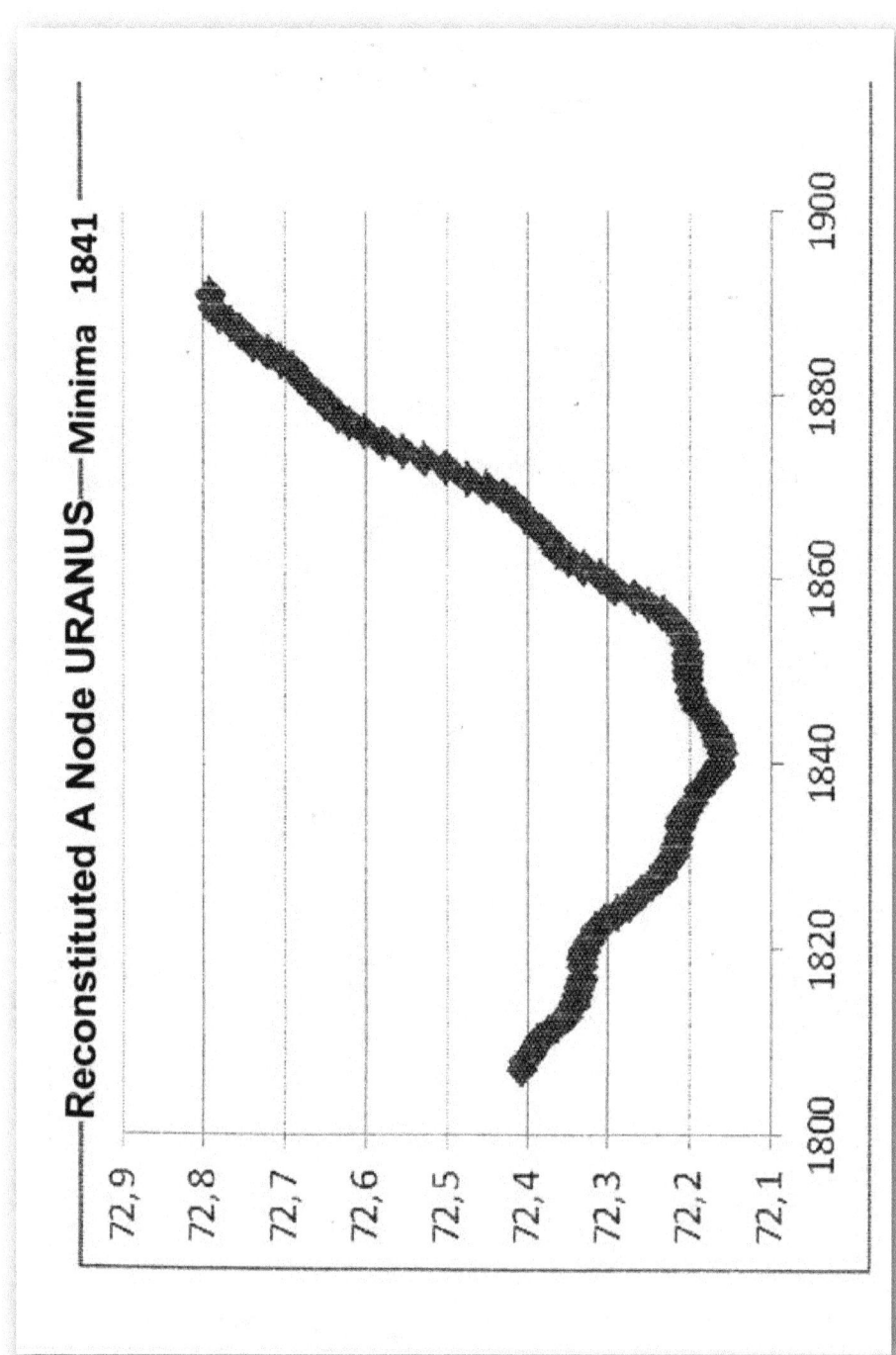

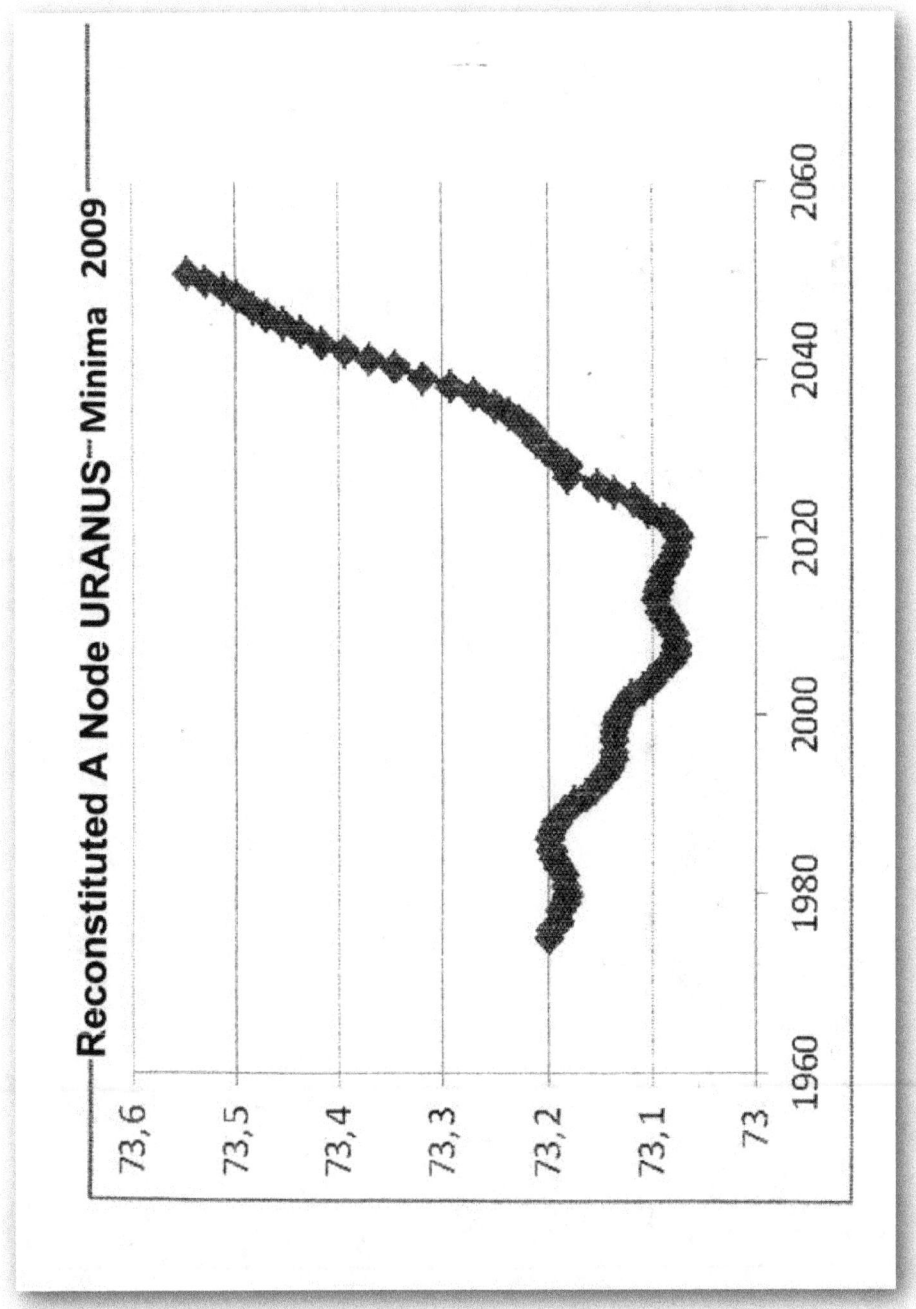

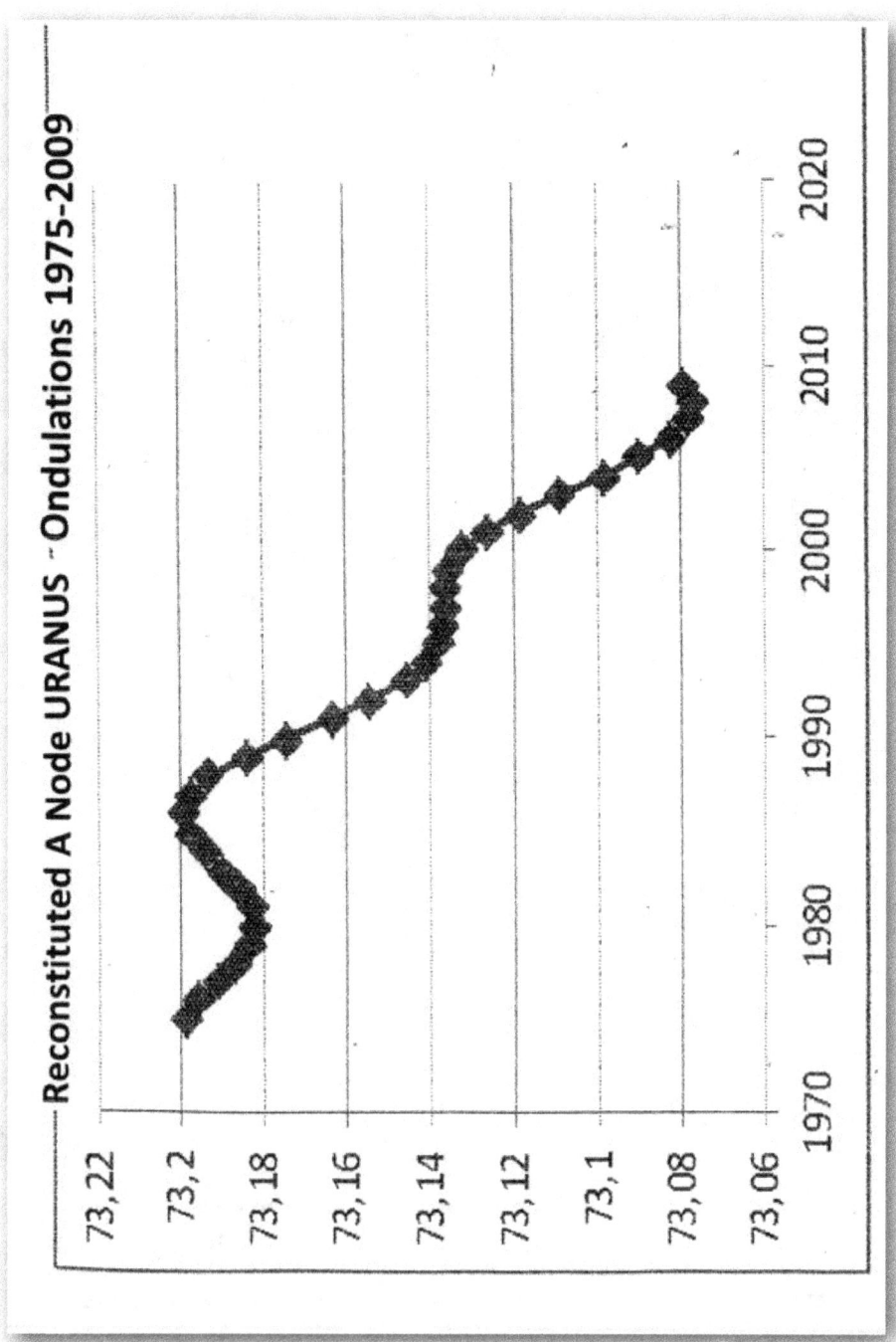

Reconstituted A Node URANUS ~ Ondulations 1975-2009

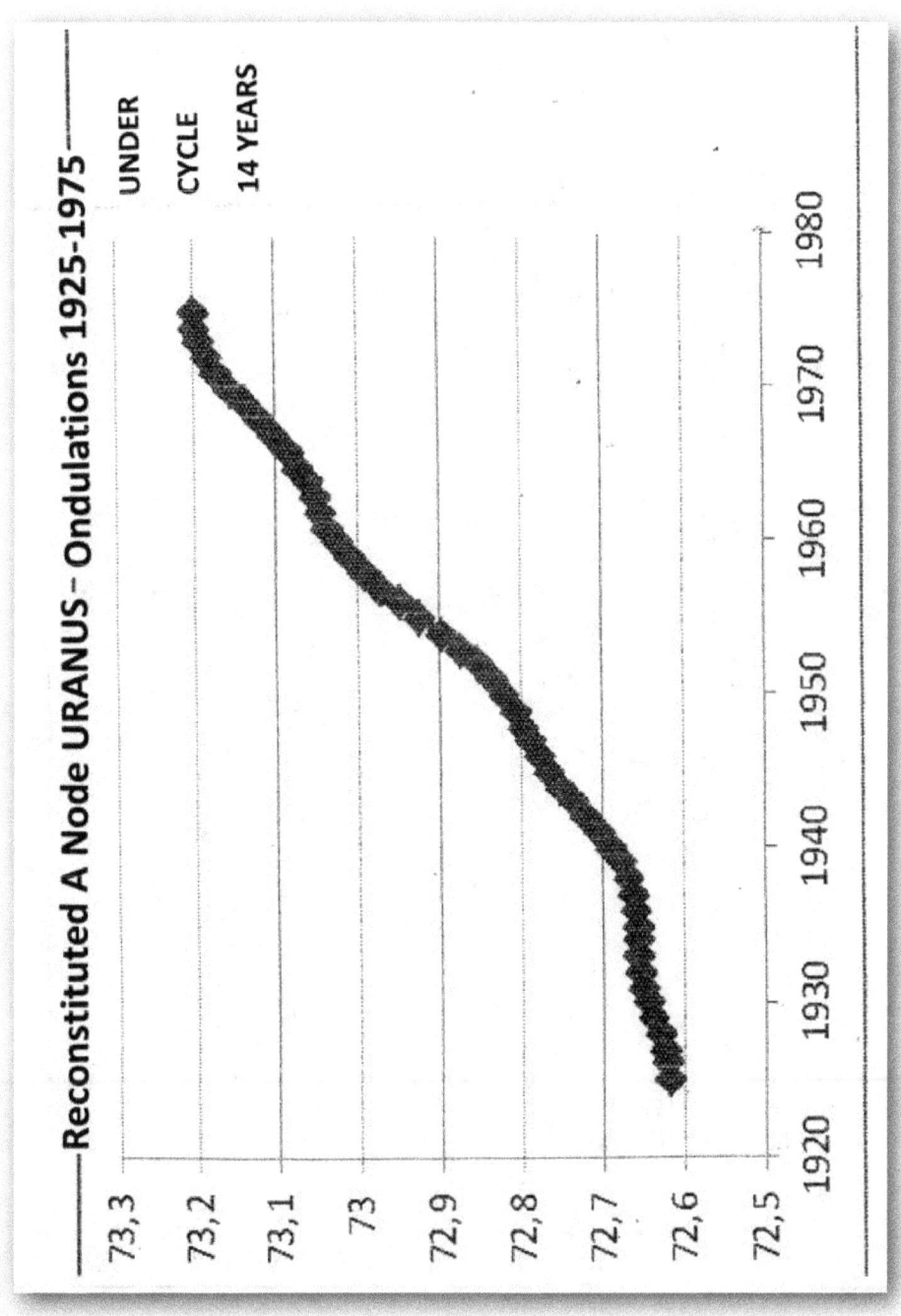

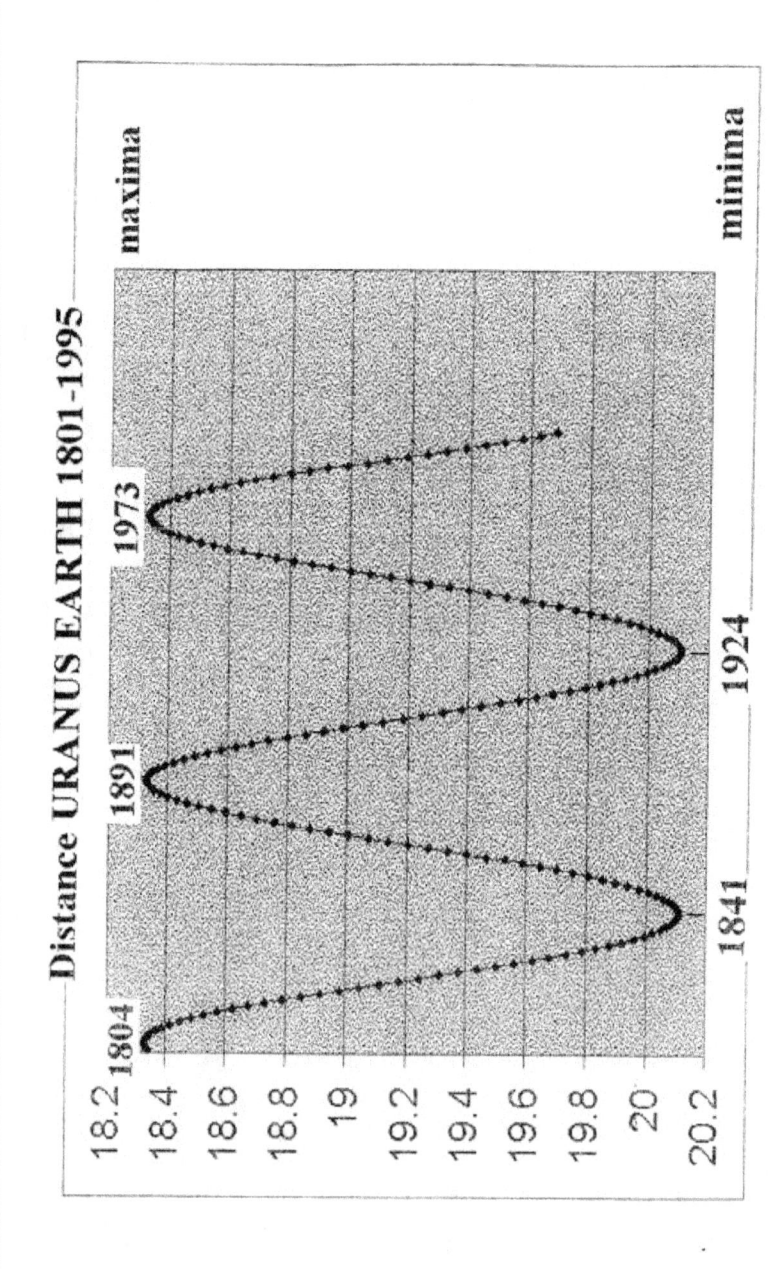

NEPTUNE

Fourth and last giant gazeous
Neptune have rings
And satellites naturals

***Calcul of the position
This planet has set
Problemes of accurate trajectory
because the distance is farther
The means used are threefold:
1)Sattelites Hipparcos, Hubble Space Telescope
2) cameras high CCD resolution
3)Observatories with mirrors
Of big diameter like Flagstaff (U.S.A.)
To complete these technical means
Recordings from 1850
Argument for complete studies of the trajectory
Global.
The recordings which allowed to
Build these bends are:
1861 to 2013 USNO (U.S.A.)
***Shape of the rising node reconstituted

For this goal i use relation LO=U+LN1 (ln1= Ω)
and i made vary LN1 (rising node)
With positions observed. One can
Get the bend reconstituted 1861-2013
The bend gotten is a quasi right.
In fact it is sinusoide very flattened
152 years in entire cycle thus 2*76 years
Taking arbitrarily the beginning cycle
In 1899:
Cycle entire 1899-2051
2 half cycle 76 years
1899-1975 half negative wave
1975-2051 half positive wave
Besides if one looks attentively
The bends reconstituted 1846-1900
And 1900-1995 one notice
Than she billow lightly.
To show in clear these ripples
With years recent i use
Bend 1925-1985
4 maxima 1935,1949,1961,1974 are noticed
13 to 14 years in average of apparition.
*What the nature of this ripple?.
These maxima coincide not with conjunctions
Nor at specifics configurations

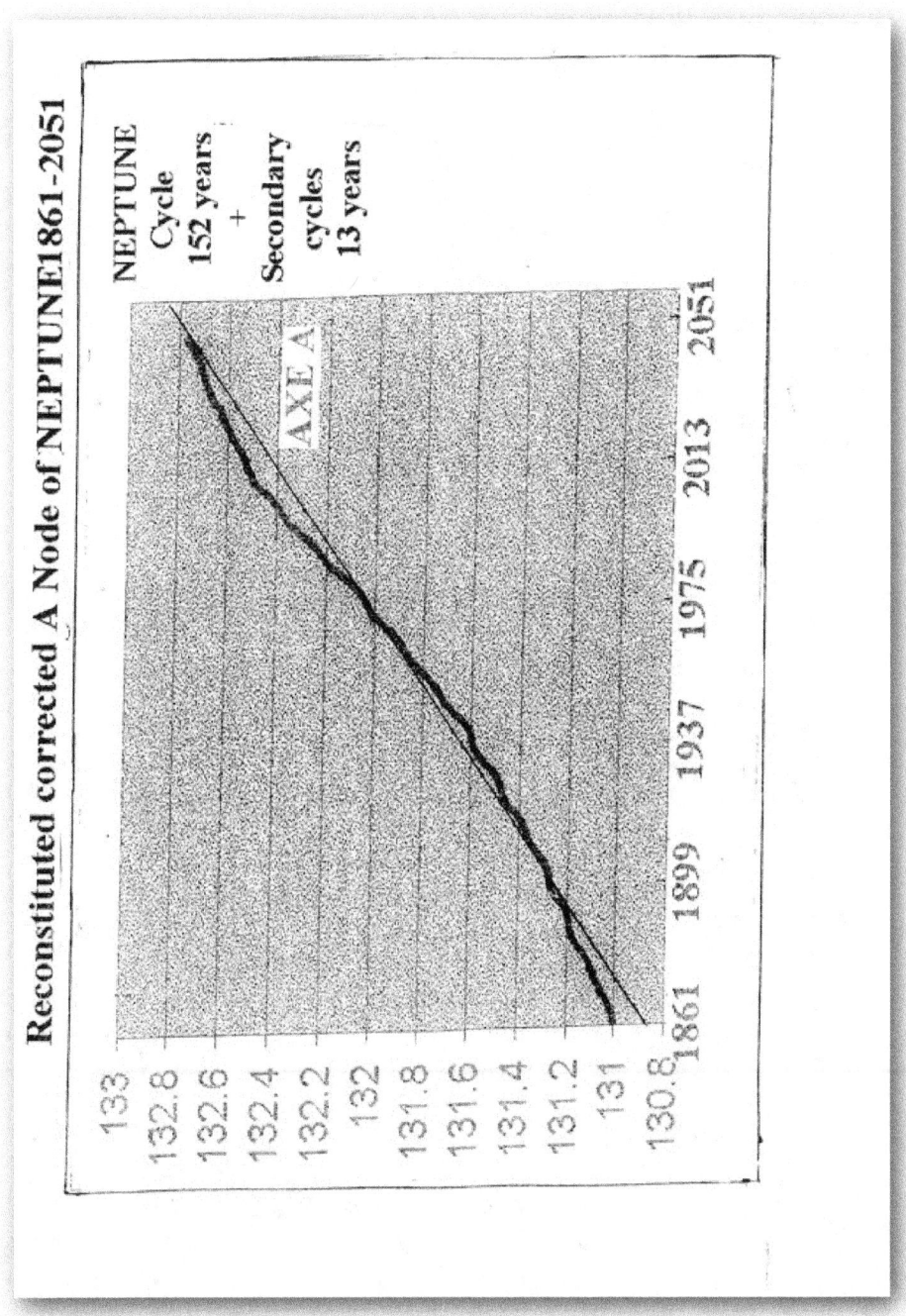

Reconstituted corrected A Node of NEPTUNE1861-2051

NEPTUNE
Cycle
152 years
+
Secondary
cycles
13 years

AXE A

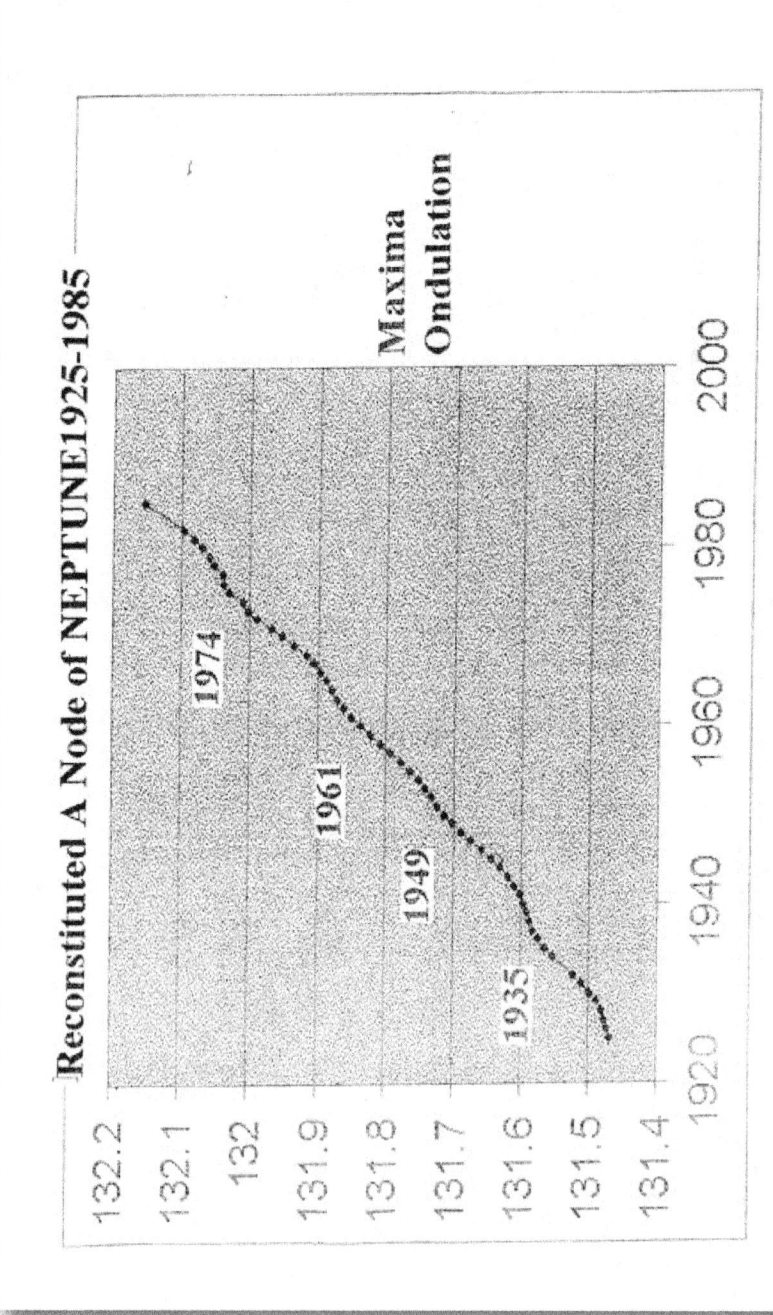

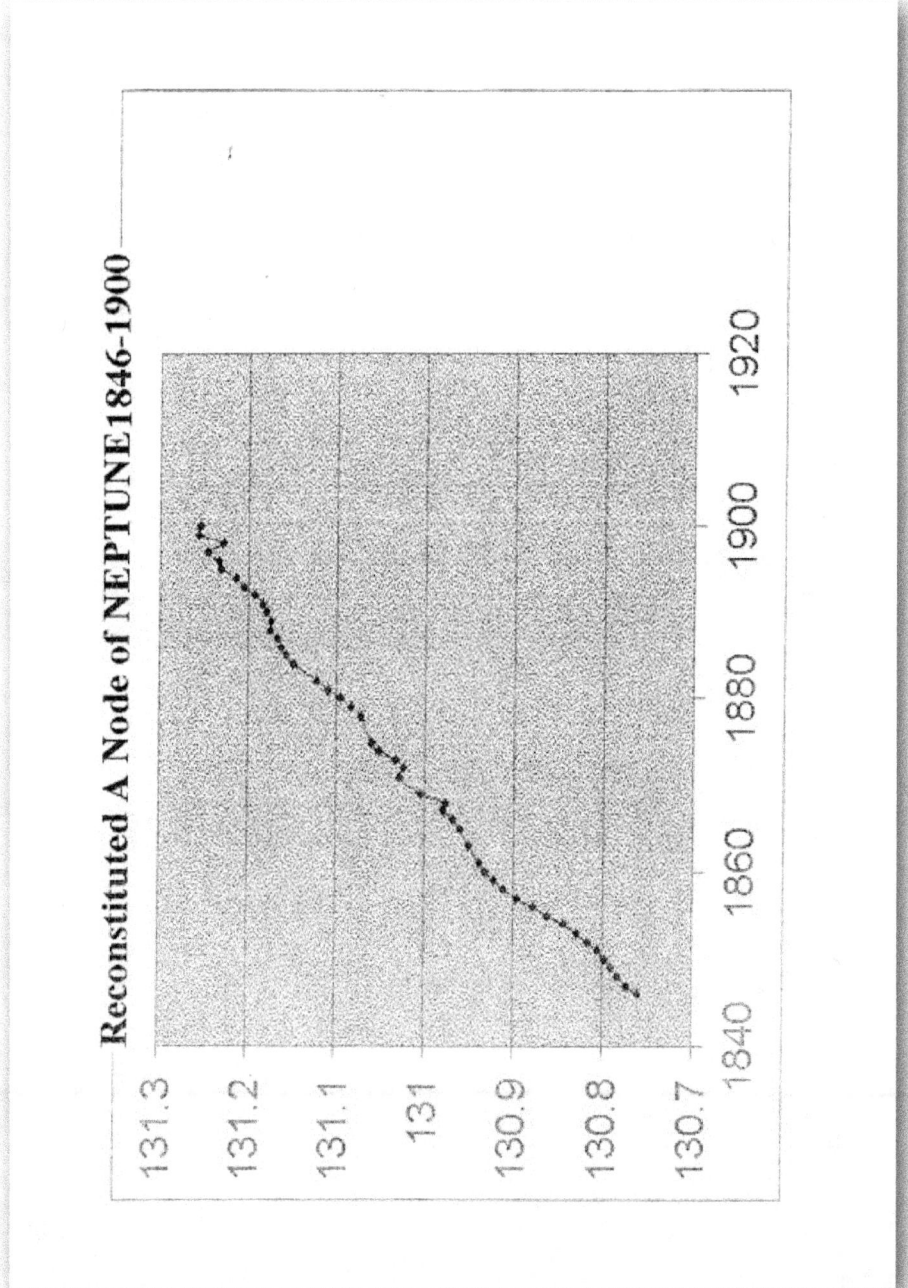

Reconstituted A Node of NEPTUNE 1846-1900

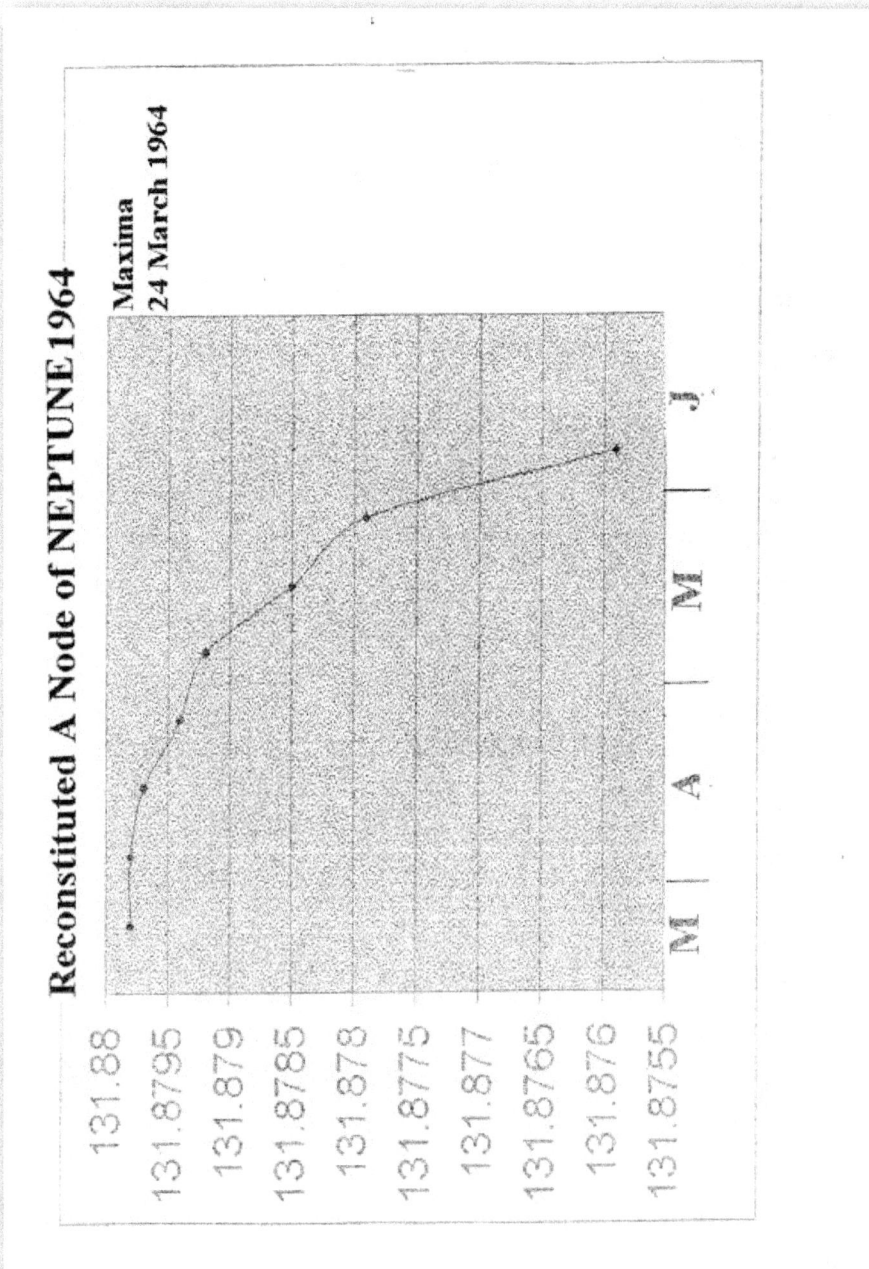

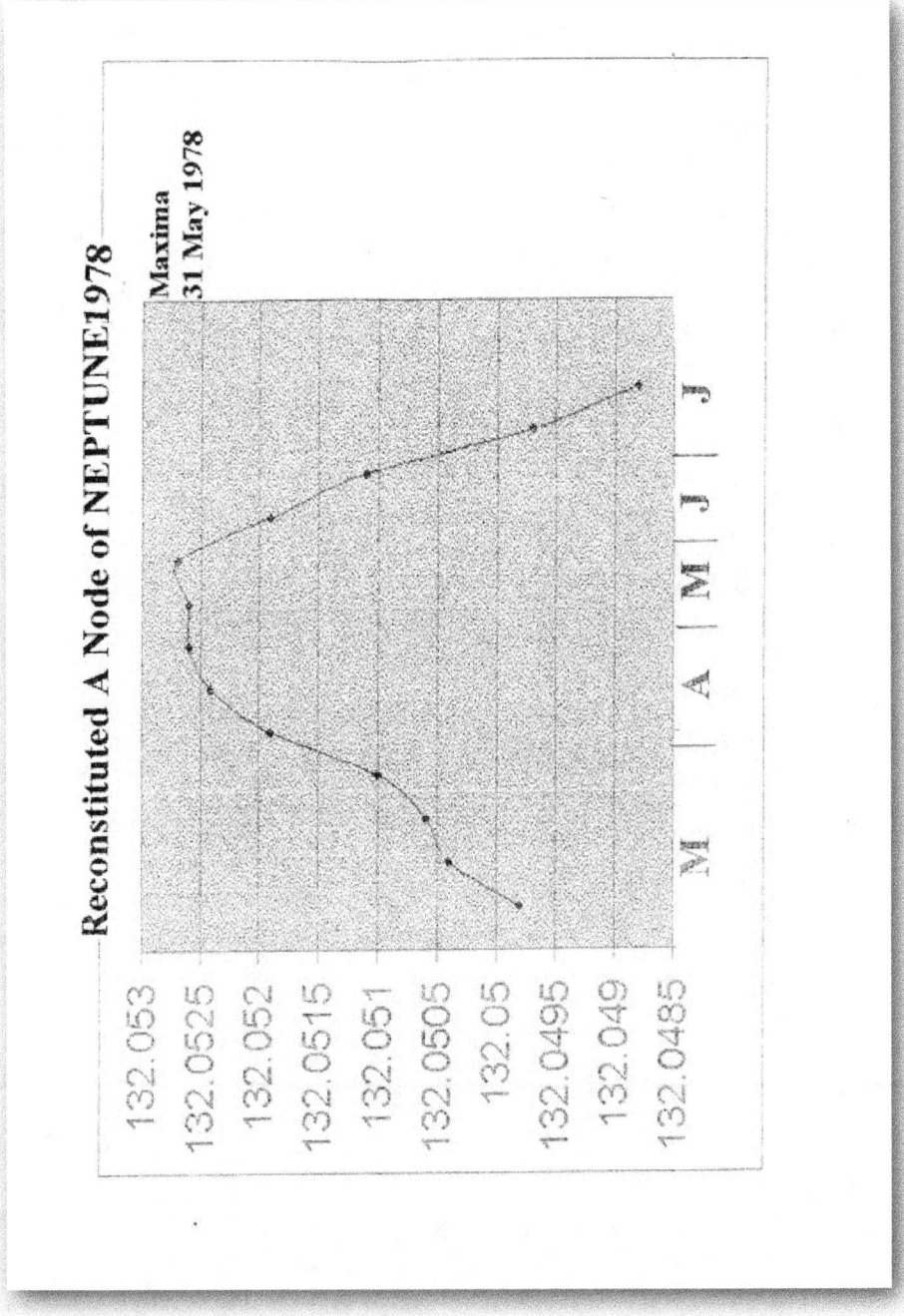

Reconstituted A Node of NEPTUNE1978

Maxima 31 May 1978

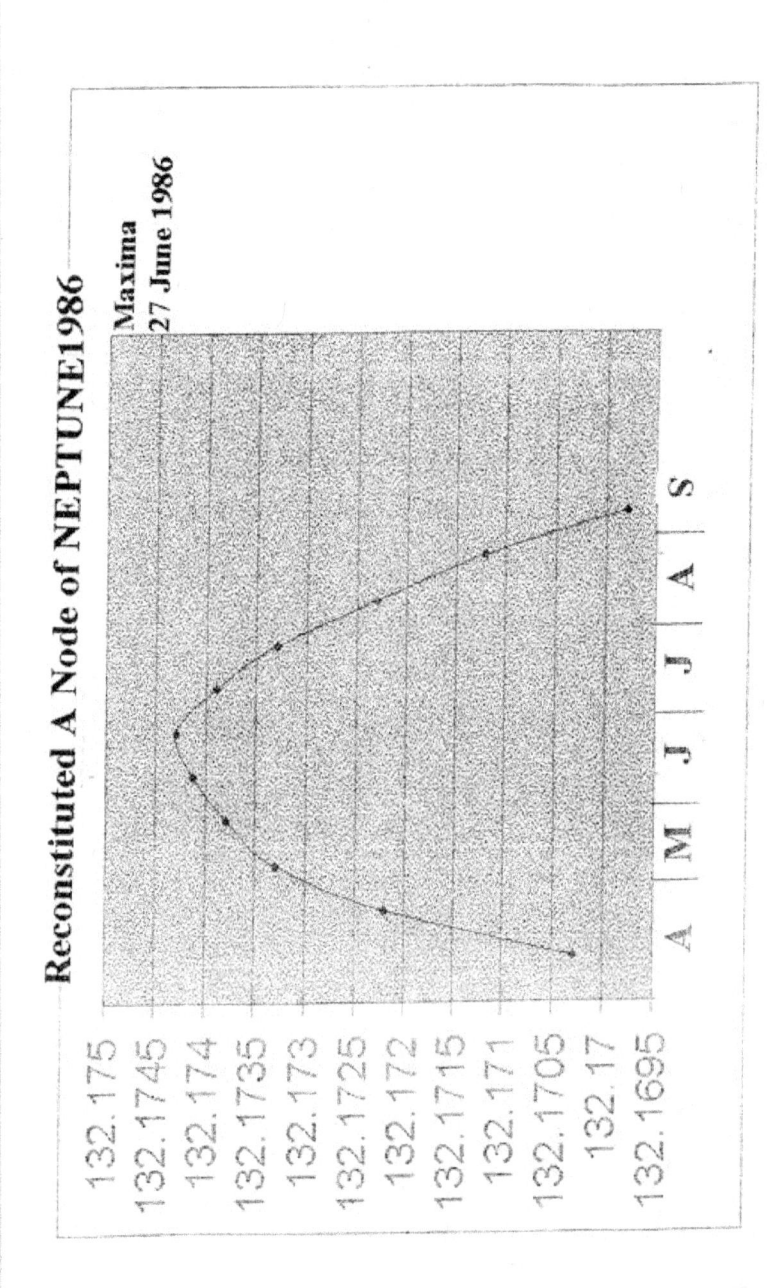

Reconstituted A Node of NEPTUNE1986

Maxima
27 June 1986

132.175
132.1745
132.174
132.1735
132.173
132.1725
132.172
132.1715
132.171
132.1705
132.17
132.1695

A   M   J   J   A   S

PLUTO

Ninth known planet today
of system solar. She was uncovered
By clyde .w. Tombaugh in 1930.
Pluto (d=2200 km) have a satellite Charon
with a big diameter d=1200 km who turns
at weak distance in 6, 39jours
More, Hubble Space Telescope
have discovered in 2005 two new
Satellites of pluto Hydra, Nix
Who carry out a complete revolution
In 38 and 25 days.
Lately a third satellite name : P4
have been discovered. It turn between Hydra and Nix.
Thus 4 naturals satellites were classified
Charon;Hydra;P4;Nix

***calcul of the position pluto
For data concerning positions older
I used in part the recordings
Incorporated in the site

'» astrometrie database planet «
From 1914 to 1993. Amongs the recordings
Some are of average accuracy
But the set stays coherent.
For the measures the most recent
I have use results of observatories
Well equiped and Hubble Space Telescope

*** reconstituted node ***

From the recordings of americans files
1914 at 2010 the bend reconstituted is gotten Pluto 1910-2010
A positive ripple is noted merely (Slowdown)
Bound in its passage at perihelie (the node move little)
The remainder of the bend has a tendency at linear.

*** Recent measures of observatories

The measures of observatories well equiped
Give differences from 0.01 to 0.02 on period
20 years for orbitals elements (J2000)
The satellite Hubble give 0.015 for a serie 1998
It is noticed that difference of measures betwneen ground
And space does not appear contrarily
At the more luminous planets.
In fact Pluto is too farther
To receive meaningful emissions from particles
Cycle solar with an increase light.

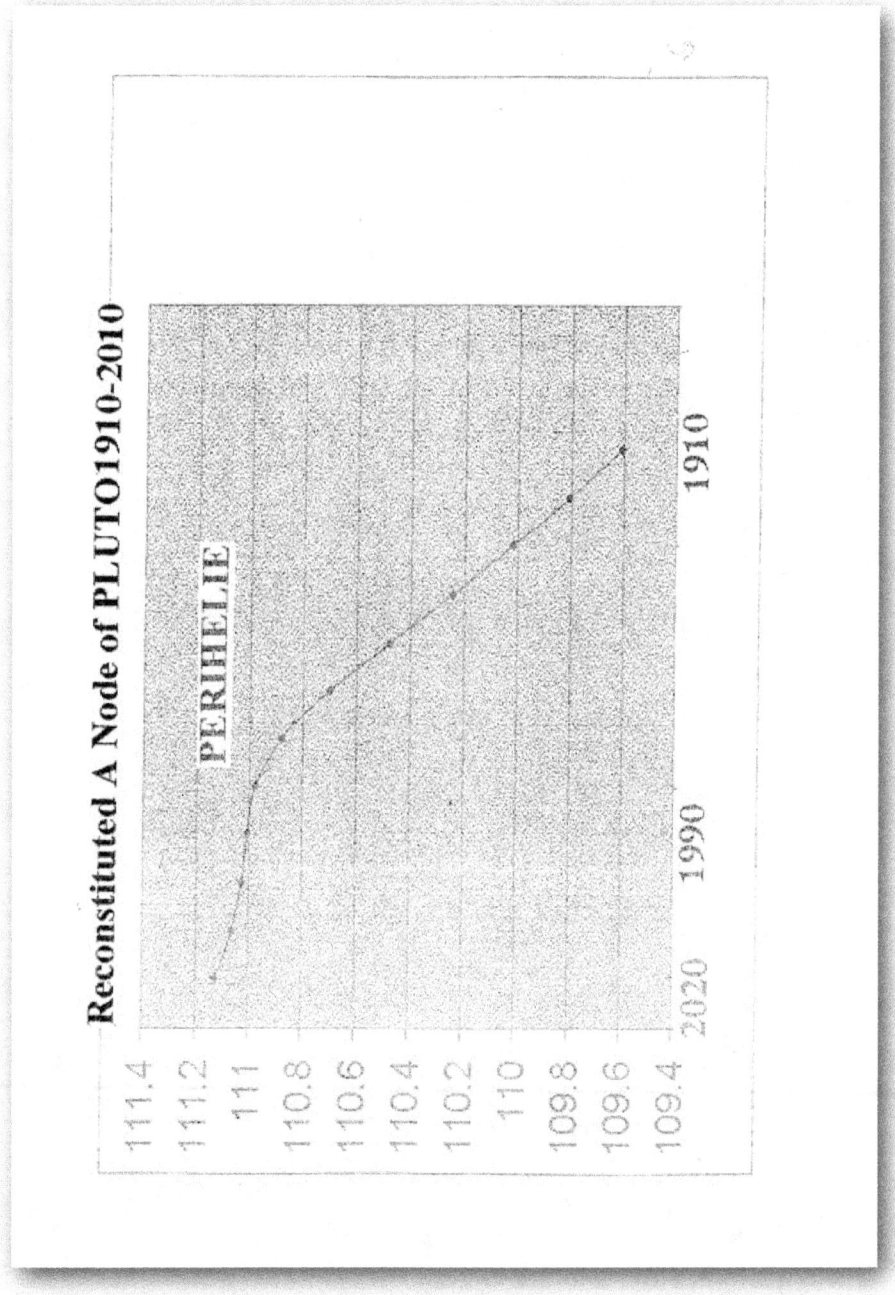

CONCLUSION

The use method of the reconstituted node
Allowed the discovery of:
The oscillation of the earthly globe with the inners
And the moon
The true cycles of the outers planets
Fundamental + secondary cycles

Accuracy measures at the ground allowed
As a whole the elaboration of computation correct at the 0.01
Degre who is the current standard
However the use of the satellite Hipparcos
On Uranus and Neptune allowed to evaluate
Difference with measures at the ground :+0.1 degre
For Neptune and >>+0.12 for Uranus
In right ascension what is high.
One must thus systemize the use of satellites
Observations on all of the planets with
series important to evaluate the differential
With the measures at the ground.

www.ingramcontent.com/pod-product-compliance
Lightning Source LLC
Chambersburg PA
CBHW070833180526
45168CB00002B/823